Total Project Control

A Practitioner's Guide to Managing
Projects as Investments

Second Edition

Industrial Innovation Series

Series Editor

Adedeji B. Badiru

Air Force Institute of Technology (AFIT) – Dayton, Ohio

Total Project Control

A Practitioner's Guide to Managing Projects as Investments

Second Edition

Stephen A. Devaux

CRC Press
Taylor & Francis Group
Boca Raton London New York

CRC Press is an imprint of the
Taylor & Francis Group, an **informa** business

CRC Press
Taylor & Francis Group
6000 Broken Sound Parkway NW, Suite 300
Boca Raton, FL 33487-2742

First issued in paperback 2020

© 2015 by Taylor & Francis Group, LLC
CRC Press is an imprint of Taylor & Francis Group, an Informa business

No claim to original U.S. Government works

ISBN-13: 978-1-4987-0677-3 (hbk)
ISBN-13: 978-0-367-78354-9 (pbk)

Library of Congress Cataloging-in-Publication Data

Devaux, Stephen A., 1949-
    Total project control : a practitioner's guide to managing projects as investments /
Stephen A. Devaux. -- Second edition.
        pages cm
    Includes bibliographical references and index.
    ISBN 978-1-4987-0677-3
    1. Project management. I. Title.

HD69.P75D48 2015
658.4'04--dc23

2014042246

Visit the Taylor & Francis Web site at
http://www.taylorandfrancis.com

and the CRC Press Web site at
http://www.crcpress.com

Dedication

To my wife, Deb,
and
to my sons, A. J. and Eric

Contents

Preface

The first edition of *Total Project Control* was published in 1999 and offered two different types of enhancement to the traditional project management approach:

1. New techniques and metrics that expanded on traditional methods, such as work breakdown structures, critical path scheduling, and resource leveling.
2. A whole new approach to projects as endeavors that are undertaken for the business value they are expected to generate.

These two areas of enhancement were intended to be closely related; the technical and metrical innovations were designed both to support and to be supported by acknowledgment of projects as investments. This meant:

- The brand new metric in CPM (critical path method) scheduling called *critical path drag* would allow project teams not only to compress their schedules, but also to measure the impact of that compression on the project's expected monetary value.
- Critical path drag's corollary metric *drag cost* would allow the project manager to not only justify the resources needed for compression, but also permit the organization to see how much the time required to execute critical activities was truly costing.
- The cost of leveling with unresolved bottlenecks (the CLUB (cost of leveling with unresolved bottlenecks)) would allow both project and functional managers not only to track the dollar impact of resource bottlenecks, but to prioritize such costs by the specific resource insufficiency in order to improve hiring strategies.
- The adaptation of the work breakdown structure (WBS) into a value breakdown structure (VBS) would allow project sponsors and customers to designate which work packages were mandatory and to place a value-added figure on optional packages. This would mean that when an optional activity was costing more (through the sum

of its drag cost and resource cost) than its value to the sponsor, that fact would become evident through its "negative value-added." A decision could then be made to either jettison the optional work or change the planned way of performing it to make it cost efficient.

- The doubled resource estimated duration (DRED) would provide a measure of the elasticity of each activity's duration in response to staffing levels and, combined with an activity's drag cost, allow project managers to assess much more quickly the relative suitability of each activity for compression through additional resources.
- Devaux's Index of Project Performance (the DIPP, an index that I introduced in the *Project Management Journal* as far back as the issue of September/October 1992) would serve as the team's primary operating metric for making decisions on the basis of project value. However, it also would guide senior management both in selecting the right projects for the portfolio and as a quick-and-easy index for checking implementation progress on the basis of a single number, one that integrates scope, schedule, cost, and risk.

The reception to *Total Project Control's* first edition among the project management community has been enormously gratifying. Project managers have recognized that the book contains not only innovative techniques, but ones that can make their jobs less nerve-racking. They can simplify, for instance, the process of finding those 10 or 12 activities in a 3,000 activity project that offer the best opportunity for recovering from schedule slippage. The project value approach simultaneously offers, as well, a way to justify the added resources that those activities need.

Since the initial publication of *Total Project Control*, theorists and practitioners have expanded on some of its ideas. Dr. Tomoichi Sato of Japan Gas Company and Tokyo University enhanced the DIPP to the *Extended DIPP* as an index of risk-based project value (RPV) for determining the best timing for risk exposure and mitigation.[1] Additionally, three different project management software companies launched packages that computed critical path drag.

Over the years since the first edition was published, I have taught several thousand practitioners the concepts, techniques, and metrics of the Total Project Control (TPC) approach—in corporate training classes, graduate courses, PMI (Project Management Institute) webinars, and meeting presentations. Many managers, after attending a seminar or reading the book, invited me to their companies to help them apply the concepts either in developing an initial plan or in recovering from a slipped schedule. These methods are all explained in this book and are current and useful tools and metrics.

However, over the years, I also have gotten feedback from some project managers that, in order to maximize the impact of this new approach, one needs to focus on expanding the understanding of corporate leaders, of their organizations, whether they are project management experts or not. For while many project management practitioners now use these techniques to deal with their schedule and budgetary dilemmas, most are not in a position to squeeze the best juice from the TPC orange—ensuring that every project is striving to maximize the business value. To accomplish this requires that an organization adopt the approach of projects as investments, and that in turn requires standardized processes, documentation, metrics, and, yes, software with a focus on the project's business value.

A couple of years ago, I was teaching at a division of a large U.S. defense contractor. As usual, the class was hungrily devouring the new ideas, recognizing their practical value. But, as so often happens, they also understood that the senior managers for whom they worked had little or no exposure to the new approach, metrics, or techniques.

"Steve," said one project manager, "these are really useful techniques. But the trouble with the overall approach of projects as investments is that you're explaining it to the master sergeants. You really need to be explaining it to the colonels and generals."

Even though senior management may be dissatisfied with all the fires that constantly break out on projects, that very firefighting prevents them from ever having the time to attend a training course about the better methodologies that have been developed over the years and that might stop many of those fires from being sparked in the first place.

However, senior managers usually do have the time to read: at lunch, on airline flights, and at home in the evening. I conceived of a different type of book, written with senior managers in mind. Instead of explaining the complex ins and outs of CPM's more arcane metrics, it would focus on redefining projects as investments that are undertaken to generate the business value that senior management wants. It would explain why so much of the current approach is counterproductive, and show how to put metrics in place to guide project teams toward the correct goals, thus making both projects and project-driven organizations much more profitable. It also would point out the sometimes immense cost of time on projects and the value of techniques to shorten the critical path.

My book, *Managing Projects as Investments: Earned Value to Business Value*, was published by CRC Press in September 2014. It summarizes the powerful new TPC techniques for optimizing project schedules, such as critical path drag, drag cost, and true cost, but without getting into the sort of detail that senior management doesn't need. It also devotes two chapters to the often misunderstood techniques of earned value management, a traditional project control method that allows senior management and the sponsor/customer to track resource usage and cost performance (and, with much less precision, schedule performance) during project execution.

The publishing of *Managing Projects as Investments* created the need for a new edition of *Total Project Control*, explaining and exploring for project management practitioners those very topics about which senior management needs only to be assured that the project team understands and is using:

- Product and project scope documents.
- The work breakdown structure (WBS) and its project investment corollary, the value breakdown structure (VBS).
- The precise algorithm for critical path analysis including drag and drag cost computation in networks with complex logical dependencies, such as start-to-start, finish-to-finish, start-to-finish, and lags.
- How to create a baseline plan that allows decision making and tracking on the basis of maximized project value.

These and other methods are the subject of this new edition of *Total Project Control*. The project manager and/or team member who becomes proficient in these techniques will not only be able to demonstrate his or her mastery, but will be able to add maximum value to the organization's project investments.

Reference

1. Sato, T. 2005. *Risk-based project value analysis: Contributed value and procurement cost.* Yokohama, Kanagawa, Japan: JGC Corporation. Online at http://www2.odn.ne.jp/scheduling/RVAnalysis/ProMAC2006%20Sato.pdf

Acknowledgments

In the years since the first edition of *Total Project Control* was published, many people have supported the new techniques that it introduced. Many have applied them in their businesses and further advanced the ideas.

Critical path drag computation in a large and complex network can be time consuming, but it is exactly the sort of problem that the computer is designed to solve. Russ Iuliano of Sumatra Development, Inc.; Vladimir Liberzon of Spider Project; and Bernard Ertl of InterPlan Systems have included critical path drag calculation in their respective project management software packages.

In the corporate world, many companies have enlisted my services in training and consulting. I trust that they have all found value in these new techniques. I would especially like to mention Tom Arsenault, Ray Brousseau, John Bugeau, Cheryl Chaput, Drew Conti, Jim Fasoli, John Gill, Mike Greene, Bob Korkuc, John Labrosse, Dan Murray, Melinda Norcross, Frank Phillips, and Ralph Titone.

Over the years, Joe Sopko, Jeff Parker, and Dr. Priscilla Glidden have all introduced the new techniques within their companies. They, along with Denise Guerin, also have been teaching these techniques for many years in their graduate and executive courses.

Dr. Tomoichi Sato of Japan Gas Company has advanced the techniques of value estimation in a whole new direction with his work and publications on risk-based project value (RPV) analysis. I thank him sincerely both for this and for the graduate classes he teaches at Tokyo University where he has incorporated many of my ideas into his lectures.

In my academic career, I have been supported by Karen von Sneidern of the University of Massachusetts/Lowell, Dr. Ken Hung of Suffolk University, Tom Carter and Sybil Smith of Brandeis University, Dave Barrett and Andrew Bennett of Olin College of Engineering, Dr. Priscilla Glidden of the University of West Indies, and the late Dr. Hans Thamhain of Bentley University.

While I have always loved teaching in any setting, my academic students have given me the greatest sense of fulfillment. All the names would be far too numerous to mention. However, some who stand out in memory include Ed Anderson, Mahesh Hegade, Dr. Timothy Hemesath, Takayuki Iida, Meena Jayaraman, Emily Ramey, and Vernon Valero of Brandeis University; Jane Maine, Alison Sullivan, and Rich Takvorian of the University of Massachusetts/Lowell; Hang Bui, Vasudevan Devarajan, Jason Edmunds, Anthony Giuffrida, Andrew Masnyj and Sukhpreet Rana of Suffolk University; and especially my students from the University of the West Indies Peter Alleyne, Neil Broome, Sharon Carter-Burke, Sidney Cox, Octavia Gibson, Rey Moe, Adrian Sinkler, and Calvin Watson. I have learned so much from those I have taught.

I have always been aware of the value that these techniques offer for those endeavors where human life is at stake. However, it was Dr. Adedeji Badiru and Major Leeann Racz of the Air Force Institute of Technology who recognized its importance and invited me to contribute a chapter on computing critical path drag and drag cost in emergency response planning to their invaluable publication from CRC Press (2013), *Handbook of Emergency Response: A Human Factors and Systems Engineering Approach*. I sincerely thank them for helping me to spread these techniques into this most crucial area.

The folks at Taylor & Francis/CRC Press have been wonderfully supportive, and I thank Cindy Carelli, Marc Gutierrez, and Kari Budyk for their help.

Finally, I'd like to thank Chip Yorke for introducing me to the writings of Daniel Kahneman, 2002 Nobel Laureate in Economics and author of *Thinking, Fast and Slow* (Farrar, Straus, and Giroux, 2013). Kahneman's book, although not explicitly about project management, explains much about the human unwillingness to invest appropriate effort in project planning, a failing about which I have often wondered.

About the author

Steve Devaux, PMP, MSPM, is president of Analytic Project Management (APM), a training and consulting company in Swampscott, Massachusetts, that he founded in 1992. APM is a Global R.E.P. (registered education provider) of the Project Management Institute (PMI). Clients include BAE Systems, Siemens, Wells Fargo, Texas Instruments, Wyeth Pharmaceuticals, iRobot, L-3 Communications, American Power Conversion, Irving Oil, and Respironics.

Devaux is the author of the book *Total Project Control: A Manager's Guide to Integrated Project Planning, Measuring, and Tracking* (John Wiley & Sons, 1999). He has worked to develop and use new approaches and metrics in project management with clients in a wide range of industries. "When the DIPP (Devaux's Index of Project Performance) Dips" was published in *Project Management Journal* in 1992 (an article that was reprinted in PMI's *Essentials of Project Control* in 1999). He has contributed chapters on his new scheduling metric, critical path drag, in two 2013 CRC Press books: *Project Management in the Oil and Gas Industries* and *Handbook of Emergency Response*. He has authored numerous articles and PMI webinars, and is a frequent speaker at PMI chapter meetings throughout the United States.

Devaux began his career at Fidelity Investments, Citicorp, and the Federal Reserve Bank of Boston. He then taught and consulted in project management at Project Software and Development, Inc. (PSDI). He has taught graduate project management courses at Suffolk University, Brandeis University, and the University of the West Indies/Barbados, and in Executive Education programs at Bentley University and University of Massachusetts/Lowell.

Introduction

Every project, no matter what the industry or work type, is a compromise among three variables: scope, time, and cost. It may be pictured as a triangle, as shown in Figure I.1.

In the fifth edition of the *PMBOK (Project Management Book of Knowledge) Guide®* (PMI, 2013), scope is divided into two separate categories. In the Glossary, *product scope* is defined as "the features and functions that characterize a product, service, or result." *Project scope*, by contrast, is "the work performed to deliver a product, service, or result with the specified features and functions." In other words, the project scope is the total amount of work that will generate, at the end of the project, a product or other deliverable that meets the specified requirements for which the project was undertaken.

Planned cost on a project is referred to as the budget, the total investment in resource usage expected to accomplish the work scope. The budget covers all manner of project expenses: labor, materials, equipment, travel, subcontractor costs, and indirect costs, such as overhead. The budget represents the intended amount of investment on the part of the sponsor or customer in order to obtain the final deliverable.

Time is the total elapsed time, from conception to completion, that it takes to perform the work. It is not work hours, which is a measure of resource usage, but rather the duration of the project.

I recommend getting used to this triangle; it will be a recurring motif in *Total Project Control*, a paradigm for thinking about all project-related matters in an integrated way.

Traditional project management deals primarily with the two diagonal sides. Indeed, project management software packages often boast of their ability to handle cost/schedule integration, i.e., to show the effect of a change in schedule on the cost and vice versa. However, what about the foundation on which the entire project rests? The scope, both product and project? Surely we need to be able to make decisions across all three variables, and to be able to see the impact of any change across all three.

Figure I.1 The project triangle.

The Development of Traditional Project Management

Most of the literature on project management specifies either 1957 or 1958 as the date of its birth. The reason these two dates are mentioned is that these are the years in which project management's greatest and most beneficial techniques were documented: CPM (critical path method) in the construction industry in 1957 and PERT (program evaluation and review technique) by the consulting company Booz Allen Hamilton on the U.S. Navy's Polaris missile program the following year. Thus, right from conception, scheduling was perceived as vital to project management. As the years passed, other project management techniques were superimposed on the original critical path methodology. In 1964, IBM enhanced CPM with the precedence diagram methodology (PDM). This allowed greater and easier scheduling flexibility. Then activity-based resource assignments (ABRAs, without the cadabra) led to resource scheduling, which led to activity-based costing (ABC, *with* the cadabra), which led to cost scheduling, which led to earned value tracking and analysis. That, basically, represents the current state of the art. Prior to the Total Project Control (TPC) methodology as described in the 1999 edition of this book, there hadn't been a major enhancement to the project manager's "bag of tricks" in almost 20 years.

Almost since the start of commercially used computers, corporations have been loading the hundreds of commercially available project management software packages onto mainframes, minicomputer networks, or microprocessors, and putting project managers in charge of projects. They have created program offices, with schedulers and accountants reporting to a program (project) manager, and, with varying degrees of discipline, holding the project manager accountable for bringing the project in on time and on budget. (The project manager defending late delivery on the grounds of insufficient resources, enhanced quality, or additional scope has had his/her work cut out, often literally.)

To this end, project management techniques have worked with differing degrees of success, in different industries, different companies, and on different projects. However, it should be noted here that, in many instances, apparent success on schedule and cost was only achieved through major sacrifices in scope, in both specific features and in quality. These sacrifices were often invisible due to the lack of an integrated scope plan, yet have often come back to haunt organizations.

However, what is undeniable is that the standard techniques of project management have led to much greater success when applied than when ignored. Plus, the more diligently these techniques have been applied, the better the chances of success. Yet, decades after CPM was codified, its detailed workings are still largely ignored by corporate America. Often, they are regarded as the bailiwick of a "plastic pocket protector-type" lead engineer who manages the project in his or her copious spare time. Senior management as often as not, takes a "hear no evil" approach: "I don't care how you do it, just get it done on time. I have more important things to worry about than these project things." If schedules slip, or fall apart, the call goes out: "Leadership! That's what we need from our project manager." Napoleon was a great leader whose soldiers would follow him anywhere. In one of history's greatest project failures, they froze to death on the Russian steppes.

Senior management is correct; leadership is needed, but, throughout the system, not just at the "project management" level. Senior management must take the steps necessary to embed project management within the organization. And, alas, more often than not, they haven't. Even if they wanted to, traditional project management doesn't show them how.

Of course, project management *should* have been a key ingredient in those corporate Total Quality Management (TQM) programs of the 1980s and early 1990s. Any company that says it is committed to TQM, yet has no procedures to ensure the vigorous application of CPM schedule optimization (see Chapters 6 and 7) to all its large- and medium-sized projects is fooling itself, and, in the process, wasting millions of dollars. Any department that does not know how much time (and money) resource bottlenecks are costing on each project, and across all projects per fiscal year, is almost certainly not staffed at its optimum level. And any CEO who does not demand a regular earned value report on the cost and schedule performance indices (CPI and SPI), not to mention the new profitability index that will be discussed in this book, the DIPP (Devaux's Index of Project Performance) should perhaps consider a job more in keeping with his or her preferred level of involvement. Because his/her organization is almost certainly wasting vast sums of money through project inefficiencies.

Project Management and the Software Industry

An oft-heard excuse for not using project management techniques is the work type or industry. This theory holds that project management may be very useful on someone else's type of project, but not on mine.

"Look," said one CIO (chief information officer), "that stuff may work when you're dealing with manufacturing or construction or something where there's a tangible product at the end and everybody knows what to do. But it just wouldn't work in software. The best programmers work by building the system as they go, and you never know until the end how long things are going to take. Stuff like critical path and all that are out the window before you even start."

This approach has led in recent years to adoption in information systems, and software development in general, of alternative processes usually referred to under the heading *agile project management*. In this process, programmers work together closely (often in pairs) and with frequent participation of the sponsor/customer through what is often termed an *iterative development* process. Software products consisting of specific screens, input fields, output reports, and other features are developed and iteratively customized over time to include the customer's more complex requirements: "That's pretty good, but we need to include these fields and those data and I'd like the font a bit larger and … ."

In general, this approach is to define the work into small packages and milestones, ensure that what has been done thus far is suitable, modify it if it isn't, and only then go onto the next enhancement.

Let me be clear. As a consultant, I directed a project in precisely this way several years ago for the information systems division of a large insurance company, and I may well use it again if ever faced with a similar scenario. The reason we used this technique (which had not yet been given the "agile" terminology) was because the sponsor and the users were unwilling or unable to focus on and outline exactly what their requirements were for the final product. As a result, stumbling along in the dark, it would have been folly to try to invest the effort in trying to create anything close to the complete final product.

There is an old project management saying: "If you don't know where you want to go, any direction is okay." Except it's not really okay, because time and effort and money will have been wasted going down the wrong road. To avoid that, we need to stop at every intersection and ask directions, and that is exactly what the agile process amounts to—a risk mitigation strategy to try to ensure that we turn back before going too far in the wrong direction.

Planning can be a painful process. The 2002 Nobel Laureate in Economics, Daniel Kahneman,[1] notes in his book, *Thinking, Fast and Slow*, that "effortful thinking" is painful. People are much more comfortable making snap judgments and reacting quickly to specific situations than doing detailed planning and computations. Additionally, committing to a specific final product and a pathway for getting there means that if we have to make changes, our original plan will have been wrong, and no one likes being demonstrably wrong. How much better, then, to make no advanced commitments until the pathway is sure.

However, in any endeavor, good planning has the ability to uncover opportunities to avoid waste and to work with greater efficiency. One simple example of this (there are many others) on a project would be to identify places where greater division of labor and additional resources might allow parallel work on different scopes that can compress the schedule. Supporters of the agile approach argue, however, that such opportunities offer too little benefit to offset the "rigidity" of the traditional "waterfall" planning approach, where detailed planning is usually conducted up to three months in advance.

While the agile approach definitely has its place when upfront scope definition is very difficult or impossible, the tendency to use it as a default is to sacrifice all the benefits that could come from demanding an adequate planning effort. Those who would define the traditional methods as rigid are showing that they simply do not understand the methodology. I will explore this misunderstanding in much greater detail in Chapter 2, with the discussion of the A–I–M approach to project planning. However, for the moment, it is sufficient to say that no competent project manager thinks that the project will go exactly according to plan. In fact, there is an old tongue-in-cheek explanation of traditional planning:

> Why does a project team and manager spend all that time and effort planning the project? Because once the baseline plan is all laid out in detail, they will have eliminated one of the million ways in which the project could actually go.

This wry explanation is obviously *not* why we plan. Instead, we plan because we *know* that the plan is going to change, but that the specific techniques of project planning—work breakdown structure, critical path analysis, resource scheduling and leveling—are all flexible formats that will allow the plan to be readily adapted in response to the need for change.

In other words, far from being rigid, the techniques of project planning are all about being agile.

However, let us not be too critical of information systems and software project management; the maturity of methods in other industries is nothing the business world should be too proud of, either. In particular, project management's greatest tool for managing and optimizing time, critical path analysis, is woefully misunderstood and underutilized. Two new CPM concepts that were introduced in the first edition of this book, critical path drag and drag cost, seem to have made a difference in some organizations by providing, for the first time, quantification and monetization for critical path work. But in most project-driven organizations, there is still much that could be done better as projects continue to take far longer than they need to take.

In organizations that aren't clearly project-driven, critical path planning is usually ignored, with project schedules often arranged through no better methodology than a project manager using the cursor with a project management software package to do nothing more systematic than drawing a Gantt chart.

Indeed, when one takes corporate America as a whole, there can be no doubt that billions of dollars are wasted every year on projects of all sizes, in all industries. A rigorous application of traditional project management methods would lead to huge improvement. Inclusion of the enhanced techniques and metrics of TPC would improve things even more.

The Software Industry and Project Management Software

One of the biggest problems in the implementation of project management continues to be the misconception that once an organization buys a software package, it will derive the full range of benefits from the project management methodology. Isn't it interesting? All of these manufacturers and engineers and construction people are relying on project management tools that a bunch of software people (who are often among the least sophisticated project management users) designed, developed, and marketed. Talk about the blind leading the lame.

There are hundreds of different project management software packages on the market. Prices run from six figures on down. Efficient utilization in many cases is limited by the lack of project management knowledge of the software designers who created them. Yet, even the most sophisticated functionality does nothing to actually manage the project; it merely computes data on the basis of information fed into it. The decisions still have to be made by a human. Wouldn't it be nice if those humans understood what the data mean, and why they are important? However, all too often, user and management ignorance initiates an investment in project management methods and software that is soon abandoned because the immediate result is not an on-time and under-budget delivery. Organizations that go as far as seting up project management offices (PMOs) all too frequently close them down in less than three years, feeling that there has been little benefit from the effort and expense.

A few years ago, I was delivering an executive-level seminar at the Virginia headquarters of a major government contractor. The topic was the implementation of a high-end project management software package.

"The trouble with project management systems," someone said, "is, garbage in, garbage out."

The executive responsible for the implementation, a highly respected individual who had been a major player in the Apollo program, immediately replied: "Oh, no, Fred, it's much worse than that. The trouble with these systems is: 'Garbage in, gospel out.'"

Not only does the project manager making decisions on the basis of data need to understand what is coming out of the system, but those people who are feeding in the data need to know what to put in, and that's asking for an awful lot in a company whose CEO thinks that project management knowledge and skills are beneath his or her notice, something that should be taken care of three or four levels down.

Integrating Projects and Profits

As long as project management is perceived as a tool specific to a given project, and as long as the parameters of projects are confined to delivering specific scope by an arbitrary deadline for an arbitrary budget (the things that the software deals with; as Marshall McLuhan pointed out, "the medium is the message"), then senior management will likely not involve itself. "We're too busy worrying about profitability to concern ourselves with inconsequentials, such as how to actually meet those deadlines we set. Just give us what we want on time and budget." Thus, project management becomes confined to the "trenches," a low-sophistication job of providing commodities for a price, and project management itself becomes a commodity.

The project manager knows that he/she is viewed as a replaceable commodity, and becomes reluctant to let senior management know when the time or resources may not be enough to meet needs. Informing senior management about slippage will lead not to additional resources, but to harassment and/or replacement. Therefore, rather than investing effort in optimizing the schedule so as to enhance profitability (the very goal, remember, that senior management is after), the project manager works to see if the plan parameters can include:

1. A padded schedule, with every activity having "wriggle room."
2. An inflated budget.
3. Nebulous work scope definition that be can trimmed invisibly to meet the deadline. (Who cares if the result is a product that breaks if you look at it? Or one whose postlaunch support costs rapidly devour the profit margin? It met the deadline and budget.)

4. A "black hole" project from which no information escapes until it's too late. (Americans may remember the surprise with which their government officials reacted to the crashing of the Healthcare.gov website in the fall of 2013. Experienced project managers just shook their heads knowingly, having seen many such "unexpected" failures over the years both in government and in the private sector.)

There is almost always a deep disconnect between the project team's goals and those of the organization. Senior management wants "profitable" projects, but is only able to quantify its wishes in terms of the two sides of the triangle that traditional project management addresses: schedule and cost. To operate smoothly, the entire organization must be:

1. driven by the single goal of multiproject profitability;
2. quantified and measured on contributions to that goal; and
3. willing to make *and accept* decisions based on quantified analysis of profitability.

These are the enhancements that the Total Project Control method offers to traditional project management, in the form of such new profitability-based data items as expected monetary value (EMV), expected project profit (EPP), the DIPP,* critical path drag, drag cost, and the cost of leveling with unresolved bottlenecks (the CLUB). The impact of their implementation can be far-reaching. Not only will good management decisions, at both the project and executive levels, be supported by quantitative data, but bad decisions will become harder to justify. ("I continued to support this development project *despite its apparently unprofitable use of resources,* because … .") The widespread corporate downsizing of recent years, for example, was frequently undertaken without a clue as to its ultimate impact on profitability. How can I say this? Because I worked with many clients who underwent downsizing, and I know that there were absolutely zero metrics in place to determine what the impact on project performance would be of reducing resources. Only TPC, as epitomized by measuring the CLUBs of different resources in terms of their impact on project value, could provide such data for a project-driven organization. Undoubtedly, many horrendous decisions have been made that seriously damaged corporations, careers, and lives.

---

* The DIPP was first formulated in the author's September 1992 issue of *Project Management Journal,* in an article titled "When the DIPP Dips: A P&L Index for Project Decisions." It has since become the basis for a software product and the entire TPC approach.

When it is demonstrably clear that:

1. Shortening a task from four weeks to two will allow product delivery to occur two weeks earlier.
2. Such shortening relies on five workers working 16-hour days, including weekends.
3. Such an early delivery would add $2 million to the product's return on investment (a not unreasonable projection).

Then, the decisions of all concerned parties (executives, project managers, labor unions, and individual workers) regarding appropriate incentives are likely to be shaped in a manner that makes such an efficiency possible. This and other aspects of Total Project Control can alter the way that project business gets done over the next few years.

Reference

1. Kahneman, D. 2011. *Thinking, fast and slow.* New York: Farrar, Straus, and Giroux, pp. 31–78.

chapter one

The nature of a project

The seed of the business world's problem with projects lies in the very essence of the project work itself. It's different from the day-to-day work of traditional businesses, different from the deal-with-each-issue-as-it-arises approach that the current corporate structure was designed to handle.

The tsunami of projects that has swamped the business world in the past 50 years has forced the creation of new strategies for dealing with this new type of work effort. Project teams, matrix structures, temps and free-lancers, worker empowerment, activity-based costing, project templates, paradigms, postmortems, and management information systems are just some of the strategies being implemented to deal with the exigencies of project work. Yet each of these innovations tends to be considered, and implemented or rejected, piecemeal. There is little understanding as to the precise characteristics of a project that each of these strategies is designed to address, nor recognition of the relationship that each has with all of the others, and, without such an understanding, the strategies will be improperly used and will likely either fail or, worse, backfire.

I still remember a meeting I attended many years ago when working for the Information Resources division of Citicorp. The systems we were installing were complex, and the process was a nightmare of obfuscation and political wrangling between the many departments involved (customer service, programming, network communications, quality assurance, documentation, and training). A new director of installations was put in place, and one of his first actions was to announce the creation of three "installation teams." Each of these would be supervised by an "installation manager." Yet, each team member would continue to report to his or her old manager. Since at the time I was totally naïve as to project management concepts, I was very confused.

As I left the meeting, I turned to my colleague, Jim, and whispered, "I don't understand. Who's our boss now, Phil or Vera?"

"Ah, Steve," said Jim sagely, "this is what's called 'matrix management.'"

"Matrix management?" This was my first exposure to the term. "What's that? How does it work?"

Jim scratched his head.

"I don't know." Then he brightened. "But I know that's what it's called."

Looking back on this episode, I can now see that the new installation director's solution, implementing project teams, was absolutely correct.

However, the way it was done, with no training and little explanation of what the problems were and how the new structure was designed to resolve them, proved disastrous. Rather than ending the turf wars, the new structure introduced yet another tier of combatants, and the result was more system bugs, greater inefficiency, slower problem resolution, and, ultimately, the resignations of some of the best employees.

The millions who have been subjected to the "matrix management solution" over the past few decades will have no trouble recognizing the above scenario. The result has been that matrix management is often identified with chaos and political mayhem. It needn't be that way. However, it will be, as long as both managers and employees fail to understand the intrinsic nature of project work, the stresses it puts on the traditional hierarchy, and the way that the political and communications issues, which invariably arise, must be addressed.

The truth is that a project is a total system, and must be dealt with in its totality to achieve the best results, and that totality requires a full comprehension of the intrinsic nature of a project.

The definition of a project

The very fact that work is unified under the penumbra of a single project has implications for the intended results and for the difficulties that are likely to be encountered. These, in turn, affect the way in which the effort should be managed. But, *why* is work unified as a project?

For many years, I would begin my management seminars with the following, commonly accepted, definition of the word. A "project," I would write on the whiteboard

1. is a group of related work tasks
2. to be performed within a definable time period
3. to meet a specific set of objectives.

I would then ask the attendees which of the three defining parameters was the one that caused companies to implement project management. They would invariably come up with the correct answer: no. 2, performed within a definable time period. Then we would move on, with me explaining how this was absolutely correct, that project management provided tools, such as critical path method (CPM) scheduling and resource leveling that were specifically designed to help managers deal with issues of deadlines and schedules.

Then one day I began to think about this. Why *was* it that project management was implemented primarily to deal with schedules? Is that "definable time period" really the most important aspect of a project? Clearly not.

In 1996, The Project Management Institute (PMI) published the first edition of the guide to the *Project Management Body of Knowledge* (*PMBOK Guide®*). Its definition of "project" has remained almost unchanged through the succeeding editions: "A temporary endeavor undertaken to create a unique product, service, or result"[1] ("result" having been added with the second edition). This definition improved things, explaining why certain work is grouped together (because of a single product, service, or outcome), and it allowed me to point to a respected source for my definition. However, it still did not get to the root of what was really bothering me: *Why do we do projects?*

The purpose of a project, its very *raison d'être*, is the specific objectives that it is undertaken to achieve—that outcome, or product, or service, or deliverable that justifies the investment of resources to that end. That can only mean that every project is an investment, the result of which is intended to have greater value than the cost of the resources.

The final deliverable of the project, whether product, service, or other result, is what the *PMBOK Guide* defines as the *product scope*, the "features and functions that characterize the product, service, or result."[2] The project management technique that specifies the details of the product scope is the work breakdown structure (WBS). This tool is helpful in identifying and controlling the product scope. In addition, the resources to perform each component of the product scope, along with their costs, gets assigned to and stored in the various work packages of the WBS.

But if the whole purpose of the project was to get the value of the scope, and the tool for defining the scope is the WBS, why is it only resource usage and cost that is stored in the WBS? Why isn't the value of the scope also stored there?

While project management's scheduling and costing techniques are *extremely* quantified, there is nothing quantified about the WBS in terms of the whole purpose for generating the scope, and, alas, what is left unquantified is extremely difficult to manage.

Schedule and cost are tangential to the main purpose of a project: to create the scope that will generate the value. Project management is the technique for managing projects; yet, its greatest techniques and its most valuable metrics concern scheduling and costing. But what about that *thing* itself, the very reason to do the project, the item the sponsor wants built or the customer wants to buy.

If the scope is the driving factor of the project investment, surely our techniques and the project management software should address this, and, if it is valuable, then surely we can estimate its value.

A project usually consists of many different components—a carburetor, bucket seats, tires, wheels, hubcaps, and various other products—all combined to create an automobile, and combining them is an important element of a new vehicle project. It's more than the simple fact that some project manager had decided to "do" them. It's even more than that the schedule must be based on a chronological relationship determined by which activities must precede others. It is primarily the fact that the value of each component is related, adding to the value of the others, so that the total value of the project "deliverable" cannot be achieved until all of the scope is finished.

With this perception, I deduced my own Total Project Control definition of a project for that first edition of this book:

1. A group of related work tasks
2. to be performed within a definable time period
3. to meet a specific set of objectives
4. whose value is interdependent

In other words, I emphasized that a project is a total system and needs to be managed as such. To do otherwise is to risk setting up a shell game, in which product features, schedules, resources, budgets, cash flow, and work performance are modified willy-nilly, on the basis of short-term benefits, without ever referencing the impact on the most important aspect of "the big picture": project value and its impact on profit.

Finally, I would come to recognize that the implications of every project being an investment (this is so obvious once it is pointed out) were usually quickly grasped by the project management community, and led to the adoption of many of the techniques from this book's first edition.

This led to the *PMBOK Guide*-inspired definition that I have been using for the past dozen years: *A project is an investment in work to create a unique product, service, or result.*

All of this might actually seem commonplace were it not for the fact that projects *are* often still managed like shell games. The "throw-it-over-the-fence" syndrome is a long-recognized problem in corporate multifunctional projects, where individual functional departments are often excluded from any role in the planning process until the partially completed product comes crashing over the cubicle wall. The waste of time and resources to do the remediation required to make the product fit the new department's specifications is enormous. The reduction in the value of the ultimate deliverable due to inadequate multifunctional input during the design phase is often even greater, yet less visible.

Project managers, the matrix organization, and multifunctional project planning teams are corporate adaptations intended to overcome a project's incompatibility with the traditional hierarchical corporate structure. They have certainly helped, but key departments in a project's work are often still ignored, or uninformed until the last minute. This is usually due to an oversight. However, much more troubling, and costly, are two aspects of the shell game that have become endemic to the very way we do projects. They are indicative of our failure to recognize the true nature of a project: that the value of the project's objectives, and the work tasks required to achieve them, are inextricably interdependent. The following is an all too common example of this problem.

A software product is being developed for the commercial market. (Actually, it could be any kind of product, but this particular problem is especially common in the software industry.) Problems have plagued the development process, and the project manager is about to be called on the carpet for late delivery and budget overruns, even though they are not her fault. Faced with this predicament, she acts quickly by

- making the user interface less intuitive;
- reducing the number of help screens; and
- cutting back on the accompanying documentation.

She is able to get the product out the door without the kind of delay or cost overrun that senior management deems unacceptable. Crowned a heroine for her quickstepping, she moves on to manage another project, and a few months later leaves the company shortly before it goes bankrupt due to the failure of its sales force to sell a software product that had rapidly acquired a reputation for user hostility, poor documentation, and a lack of adequate telephone support.

The first aspect of the shell game in the above case is quite easy to see. The project manager was responsible, and the project had been budgeted for a user-friendly interface, and both online and printed documentation. However, when push came to shove, the time and money for those features got pushed over to the software development problems, and the job of dealing with the inadequate interface and documentation was shoved over to two areas that the company *did not identify as part of the project*: telephone support and, most importantly, sales and marketing.

Both of these areas are almost invariably omitted from project calculations and project accounting. This provides a glorious opportunity to organize the shell game, as well as the chance to obfuscate the shortcomings of any of the involved departments or individuals through pointless finger-pointing (pointless because there is seldom a clear paper trail to show where the fault lies). Instead, everyone can happily blame someone else, while both the product and the company lose money. But that's not a project manager's issue, is it? After all, her job was just to bring the project in on time and within budget, right?

This brings us to the second aspect of the shell game, one which occurs in almost every corporate project performed in the United States: There is a *total* disconnect between the performance of the work and the process by which project's value is generated (whether that process is called marketing, sales, or delivery). The project manager's mandate is to complete the project work, to specification, in a timely and economical fashion; it is *not* to increase the profit generated by the deliverable. That is marketing's problem. The metrics that the project manager works to are project duration and budgetary costs. Indeed, there *is* no metric, in traditional project management, for product profitability.

This, more than anything else, is what the "total" in Total Project Control (TPC) means: The scope as represented by the deliverable is the most important part of the project. It is the motivating force behind the project, the "thing" that is desired by the all too often forgotten customer, and, as such, it must be tied, *in its totality*, to the traditional project management responsibilities: schedule and cost. In the TPC methodology, this is accomplished through a new data metric, the DIPP (Devaux's Index of Project Performance), as defined in my 1992 article "When the DIPP Dips" in the September/October issue of *Project Management Journal*. Precisely how the DIPP works is described in Chapter 2. Suffice it to say that the DIPP is a tracking index for project profitability.

In other words, in the TPC approach, the true nature of the project, as a collection of interdependent work items that provide value, has to be fully comprehended so that the project can be managed accordingly.

The multiproject portfolio

The purpose of the individual project is to create a product, service, or result that will allow (often through a larger program) for maximum value generation above cost. So, too, the purpose of all the projects within the organization's multiproject portfolio is to generate value, to enlarge the company's bottom line. The total value minus total costs of all of the corporate projects often represents the profit margin of project- or product-driven organizations. To this end, project information must be visible at a summary level in order to ensure that decisions made on one project are not so detrimental to another project (or projects) as to reduce the profitability of the overall portfolio.

Frequently, two projects within the same organization may be in need of precisely the same resource at the same time. This is the problem that the multiproject resource leveling function of traditional project management software was developed to handle. Unfortunately, the traditional methodology has no means of dealing with project value. Therefore, even when the software tells the users that there is a problem, and even when it alleviates that problem to the best of the software algorithm's resource leveling capability, *the software may actually be doing more harm than good*, blindly resolving the bottleneck in a way that reduces the organization's profits. Because there are no metrics, the revenues disappear without a trace.

There are thousands of corporate organizations that depend on projects for more than 90 percent of their revenues. Yet, other than intuitively,

Table 1.1 Portfolio Report Displaying Traditional Project Management Data

1	2	3	4 (2–3)	5	6	7
Proj. name	EMV (000)	Budget	Start EPP	Plan compl.	Curr. compl.	Cost ETC
A	$1,000	$800	$200	1-Aug	1-Aug	$200
B	$2,000	$1,500	$500	1-Oct	1-Oct	$1,000
C	$5,000	$3,000	$2,000	25-Nov	25-Nov	$2,000
D	$10,000	$8,700	$1,300	30-Jan	30-Jan	$3,000
Portfolio	$18,000	$14,000	$4,000	30-Jan	30-Jan	$6,400

they have no way of tying the projects they do to their profits. That means that they cannot measure precisely each project's expected profits, and they have only guesswork on which to base such decisions, such as resource targeting, hiring, staff downsizing, project scope reductions, and critical path crashing. There is no data to help senior management maximize profits on a multiproject basis. This can lead to such unfortunate situations as the following:

Imagine that a product development company has a portfolio of four products: A, B, C, and D. Each has a planned delivery date as shown in Table 1.1, and each is expected to generate revenues. Although expected monetary value (EMV) often includes value other than simple revenues, here we have pretended the two terms are identical and shown the expected revenues in the column "EMV." The EPP column indicates the expected project profit, the difference between the EMV and the budget for each project. (It is important to understand that although some department managers and executives may have the EMV and EPP data, this information is rarely given to project managers.)

The "Cost ETC" column shows the cost estimate-to-complete, or the amount of money budgeted to bring each project from its current status to completion.

These data are the absolute minimum that the senior manager in charge of these four projects should have available to him. If the organization is doing a good job of traditional project management, the senior manager, in fact, should have this information. (Whether or not revenue information is up to date, or whether it is ever juxtaposed against the project cost data is another issue.)

However, even these data are inadequate, and to see why, let us now imagine that another product is proposed. Let us further imagine this new product is expected to be extremely profitable, generating fully 67 percent of the revenues of the other four products *combined*. When a detailed plan of the project (Project E) to create the new product is assembled, its budget

Table 1.2 Traditional Project Management Data for the New Project

1	2	3	4 (2-3)	5	6	7
Proj. name	EMV (000)	Budget	Start EPP	Plan compl.	Curr. compl.	Cost ETC
E	$12,000	$3,000	$9,000	10-Feb		$3,000

shows the revenues generated will be 400 percent of the anticipated costs (Table 1.2).

What vice president could resist such an opportunity? Who could possibly reject such a profitable new product? And, yet, there is not enough information here to make an informed decision, one way or another, about undertaking Project E. Why? Because nowhere are there data that show how this new project, and its need for resources, will affect the schedules, the delivery dates, and, thus, the expected revenues of the other four projects. The only way the senior manager could undertake Project E with confidence would be by staffing it with an entirely new set of resources. (This may indeed be the best decision, but, in the real world, how often does this happen? Almost invariably, the same pool of resources are required simply to shoulder the burden of the additional work and perform whatever miracles are necessary.) Without access to expected monetary value data for each and every project, the vice president will almost certainly elect to take on what seems like a profitable new project, and when the organization's profits decrease, it will be blamed on bad luck, or poor project management, rather than the lack of adequate data.

What data should the vice president have available? He should have, specifically, data about the expected monetary value of each project and for the portfolio as a whole, as well as an index that allows tracking such value and seeing how it may change in response to changes in the project plans. This index is the Tracking DIPP.

The Tracking DIPP: Setting the baseline for expected project profitability

In 1990, at a time when I had been teaching project management theory for corporations for about two years, I found myself reading many project management articles on the subject of how to know when an organization should continue funding a troubled project and when it should terminate it. Shortly after playing in a backgammon tournament, the thought occurred to me that the continue-or-terminate decision point on a project is quite similar to such decisions in gambling games—when things aren't going well, but you still have a chance, should you hold 'em or fold 'em? In simplest terms, it comes down to two issues:

- How much value is there in the situation if we continue to invest?
- How much more are we likely to need to invest before the situation is resolved?

This insight spurred me to develop a formula to analyze the variables in such a project situation. I called the formula the DIPP (Devaux's Index of Project Performance).

The DIPP formula was stated as follows:

DIPP = (TPCM – OC – CW) ÷ (Cost ETC – PTC)

where TPCM is the total project contribution margin (the equivalent of EMV), OC is opportunity costs, CW is cannibalization worth (or salvage value), Cost ETC is the cost estimate-to-complete (factoring out sunk costs) if we continue funding, PTC is project termination costs if we elect to shut it down, and all elements are adjusted for risk and discounted by a common time cost of money factor. If the numerator is greater than the denominator, the DIPP will be greater than 1.0, suggesting that there will be greater value in completing the project than in canceling it. If the DIPP is less than 1.0, that suggests that the project should be canceled unless the plan for the rest of the project can be changed so as to generate a DIPP for the remaining work that is above 1.0.

The *PMJ* article caused a bit of a stir in some circles when it was written and, in 1999, it was reprinted as a chapter in Pinto and Trailer's compilation of theoretical articles titled *Essentials of Project Control*,[3] published by the Project Management Institute.

However, while the focus of the article was on the continue-or-terminate decision, it was a few years before it dawned on me that, in fact, a simplified version of the DIPP formula, focusing on project value, the impact on that value of schedule changes, and the cost estimate-to-complete would be of great utility for analyzing project decisions across all sides of the Triple Constraint Model:

1. Increases or decreases in scope, resulting in cost and/or schedule changes
2. Changes in resources, resulting in scope and/or schedule changes
3. Risk mitigation strategies, potentially impacting scope, schedule, and cost

In the 1999 edition of this *Total Project Control* book, I proposed what I have since called the *Simple DIPP* or the *Tracking DIPP*. The Tracking

DIPP is designed to measure and integrate the three sides of the Triple Constraint Model into a single index, the essence of which is the business value of the project investment. The formula is

DIPP = ($EMV ± $acceleration/delay) ÷ $Cost ETC

The $EMV is estimated based on the project finishing on a specific date and the $acceleration/delay is the increase or decrease in value if the date of completion changes.

In the baseline version of this formula (which should be saved at the start of the project in order to track actuals against it), both values in the numerator are assumed to remain fairly constant throughout the project, while the Cost ETC is planned to decrease as work is completed. If either scope, completion date, or Cost ETC changes, the impact should be visible in an actual DIPP that is different from what the planned DIPP was predicted to be at that point.

Tracking EMV and the DIPP at the portfolio level

If one wishes to see what the impact of changes is in terms of the business value of both projects and portfolio, then these project data must be available and analyzed not just at the project level, but also across the portfolio. The senior manager should have the following TPC Portfolio Summary Report available to him for the current portfolio, as shown in Table 1.3, and similar data for the new project, Project E, as shown in Table 1.4.

The TPC reports do not treat EMV as a constant, but instead show how each project's value will be impacted by delay or acceleration due to resources having to be shared with Project E. Additionally, they show the Tracking DIPP as the index of profitability per cost dollar. Project E has not yet been scheduled and so has no Current Completion Date, Planned DIPP, or Current DIPP. However, Project E has a potential Starting DIPP of 4.0 if we can complete it by February 10, which will decrease if the project is delayed due to an EMV reduction of 20 percent or $2.4 million per week. Expected monetary value is almost invariably dependent on completion date (whether impacted by market window, contractual penalty clause, net present value, or simply delay in the time at which the benefits of the project work start to accrue). There likely will be ripples caused by dropping Project E into the resource pool shared by the other four projects. These must be identified and analyzed, not just in terms of their effect on the completion dates, but in terms of the effects of such delays on the EMVs of *each* project and *all* the projects.

A good project management software package, with resource leveling capability, is indispensable. The vice president needs to be able to perform resource leveling on a multiproject basis, generating schedules for all of the projects that take into account the constraints of limited resources.

Table 1.3 The TPC Portfolio Summary Report for Four Projects

1	2	3	4 (2÷3)	5	6	7	8	9	10 (2÷3)	11 (2÷7)	12 (2÷7)
Proj. name	Start EMV (000)	Budget	Start EPP	Plan compl.	Curr. compl.	Cost ETC	% Loss per week late	% Gain per week early	Starting DIPP	Planned DIPP	Current DIPP
A	$1,000	$800	$200	1-Aug	1-Aug	$200	5%	5%	1.2	5.0	5.0
B	$2,000	$1,500	$500	1-Oct	1-Oct	$1,000	10%	5%	1.3	2.0	2.0
C	$5,000	$3,000	$2,000	25-Nov	25-Nov	$2,000	20%	2%	1.7	2.5	2.5
D	$10,000	$8,700	$1,300	30-Jan	30-Jan	$3,000	10%	5%	1.1	3.3	3.3
Portfolio	$18,000	$14,000	$4,000			$6,200			1.3	2.9	2.9

Table 1.4 The TPC Summary Report for Project E

1	2	3	4 (2–3)	5	6	7	8	9	10 (2÷3)
							% Loss per week late	% Gain per week early	
Proj. name	EMV (000)	Budget	Start EPP	Plan compl.	Curr. compl.	Cost ETC			Starting DIPP
E	$12,000	$3,000	$9,000	10-Feb		$3,000	20%	5%	4.0

Such a software package will produce schedules that reflect the impact of Project E, showing that each of the other projects' completion dates will have to be delayed. But what even a good software package, offering traditional functionality, will produce is a schedule that fails to take into account, or even reflect, the impact of such delays on the organization's revenues and/or profit. Is not that precisely the information that the portfolio manager desperately seeks?

Table 1.5 shows what might happen when resources for the five projects are scheduled through the resource leveler, but with Project E's completion date fixed at February 10.

The inclusion of Project E with a fixed completion date to guarantee its value has so delayed the other four projects that both the DIPP and, more importantly, the net profit of the organization across the portfolio have shrunk. Yes, Project E offers an EMV of $12 million based on a cost of $3 million, or $9 million of profit; however, once it is scheduled, it also causes resource bottlenecks that (if only we had the data) are projected to cause delays on the other four projects' critical paths that will lead to a total reduction in value of $10 million. Instead of spending $14 million to make $18 million (as shown in columns 3 and 2 in Table 1.3), we will be spending $17 million to make $20 million; we have reduced our portfolio profit by $1 million.

Resource leveling algorithms are programmed to level (i.e., smooth out) bottlenecks in the same way that projects are typically managed in traditional project management—by seeking to reduce costs or shorten project durations. However, neither of these should be the main goal of project management. The main goal should be to increase profits. But traditional software has no functionality that incorporates such data. This becomes even more costly when making resourcing decisions on a multiproject basis.

If the full range of TPC data is available, the senior manager can at least be aware of the negative impact of Project E, and perhaps manually change the parameters to try to generate a more profitable portfolio profile. Perhaps E should be discarded, because its delay penalty will be so great. Perhaps C should be terminated and its resources cannibalized for A, B, and D. Perhaps approving the hiring of a few additional key resources can change the entire context, increasing ETC, but also boosting EMVs sufficiently to increase net profit. Or perhaps a change of project priorities in

Table 1.5 The TPC Portfolio Summary Report for All Five Projects, with Project E's Completion Date Fixed

1	2	3	4 (2-3)	5	6	7	8	9	10	11 (2÷3)	12 (2÷7)	13 (10÷7)	14 (13÷12)	15 (10-3)
	Start						% Loss	% Gain	Current					
	EMV		Start	Plan	Curr.	Cost	per week	per week	EMV	Starting	Planned	Current		Current
Proj. name	(000)	Budget	EPP	compl.	compl.	ETC	late	early	(000)	DIPP	DIPP	DIPP	DPI	EPP
E	$12,000	$3,000	$9,000	10-Feb	10-Feb	$3,000	20%	5%	$12,000	4.0	4.0	4.0	1.00	$9,000
A	$1,000	$800	$200	1-Aug	29-Aug	$200	5%	5%	$800	1.2	5.0	4.0	0.80	$0
B	$2,000	$1,500	$500	1-Oct	29-Oct	$1,000	10%	5%	$1,200	1.3	1.7	1.2	0.71	–$300
C	$5,000	$3,000	$2,000	25-Nov	23-Dec	$2,000	20%	2%	$1,000	1.7	2.5	0.5	0.20	–$2,000
D	$10,000	$8,700	$1,300	30-Jan	6-Mar	$3,000	10%	5%	$5,000	1.1	3.3	1.6	0.48	–$3,700
Portfolio	$30,000	$17,000	$13,000			$9,200			$20,000	1.3	2.9	2.0	0.69	$3,000

Table 1.6 The TPC Portfolio Summary Report for All Five Projects, Resource Leveled to Optimize Profit

1	2	3	4 (2-3)	5	6	7	8	9	10	11 (2÷3)	12 (2÷7)	13 (10÷7)	14 (13÷12)	15 (10-3)
	Start						% Loss	% Gain	Current					
Proj. name	EMV (000)	Budget	Start EPP	Plan compl.	Curr. compl.	Cost ETC	per week late	per week early	EMV (000)	Starting DIPP	Planned DIPP	Current DIPP	DPI	Current EPP
E	$12,000	$3,000	$9,000	10-Feb	13-Jan	$3,000	20%	5%	$14,400	4.0	4.0	4.1	1.03	$10,900
A	$1,000	$800	$200	1-Aug	12-Sep	$200	5%	5%	$700	1.2	5.0	3.5	0.70	–$100
B	$2,000	$1,500	$500	1-Oct	15-Oct	$1,000	10%	5%	$1,600	1.3	1.7	1.6	0.94	$100
C	$5,000	$3,000	$2,000	25-Nov	25-Nov	$2,000	20%	2%	$1,000	1.7	2.5	2.5	1.00	$2,000
D	$10,000	$8,700	$1,300	30-Jan	27-Feb	$3,000	10%	5%	$6,000	1.1	3.3	2.0	0.61	–$1,700
Portfolio	$30,000	$17,000	$13,000			$9,200			$23,700	1.3	2.9	2.6	0.90	$6,700

leveling the resources can help. All of these should be tried, with the TPC Summary Portfolio Report reflecting the results so that the best solution can be adopted. Table 1.6 shows what one solution might be.

The new TPC Portfolio Summary Report shows that this particular targeting of resources will generate a schedule with an extra $500,000 to be spent on Project E, not only alleviating some of the bottlenecks, but actually pulling E's schedule *earlier* than its original target date in order to take advantage of its generous acceleration premium. The EMVs of Projects B, C, and D all go up slightly from the scenario in Table 1.5 due to bottleneck alleviation. The almost-completed Project A is delayed a little more, although its current DIPP being above 1.0 indicates we still should continue to fund it, and the net profit of the portfolio goes back up to $6.7 million from its $3 million in Table 1.5. With the aid of the TPC portfolio data, we have spent an extra $500,000 to increase the value of our portfolio by $3.7 million.

Even though such modifications may have to be performed manually, the benefits to the bottom line are huge. Ultimately, this technique should allow resource leveling algorithms to level resources for "right-sized" staffing levels and maximum profit.

Conclusion

- The purpose of a project is not to be short, or inexpensive, but to generate maximum business value above cost. It should be managed in such a way as to maximize that profit.
- All the work, and all aspects, of the project that impact its profit should be analyzed together, in an integrated way that shows the effect of the various alternatives on the project profit.
- Each project that is managed in a context with other projects should be analyzed in an integrated way that shows the effects of each (ostensibly internal) project decision on all the other projects, and, specifically, on the multiproject profit.

Insofar as projects are managed without regard to profit, bad (profit-reducing) decisions will be made, both randomly and systematically, throughout the organization.

References

1. Project Management Institute. 2013. *A guide to the project management body of knowledge.* Newtown Square, PA: PMI, p. 553.
2. Ibid, p. 552.
3. Devaux, S. A. 1999. When the DIPP dips. In *Essentials of project control*, eds. J. K. Pinto and J. W. Trailer (pp. 129–141). Newtown Square, PA: Project Management Institute.

chapter two

An overview of TPC planning

How do projects get planned? The sad truth is that, far too often, they barely get planned at all. Like Topsy from *Uncle Tom's Cabin*, a terrifying number of projects, in all industries, "just growed." This is not the case in all cultures. In Japan, for instance, it would be unthinkable to start work on a major corporate project without spending many long hours on team meetings to discuss schedules, scope, resources, and other issues. However, U.S. culture is very different. We worship the television hero who, often through his own lack of foresight, finds himself surrounded and imperiled. A blazing .44 Magnum is a lot more exciting than avoidance of danger by proper anticipation, and the project manager who acquires the reputation as a wizard at firefighting tends to get more credit than the one who quietly and efficiently removes all combustibles before the conflagration.

Projects of any significant size and/or complexity that have not been planned rapidly descend into chaos. Thus, in a multiproject organization, they often bring other projects crashing down as well.

For the unplanned project, there are at least six pitfalls that can be enumerated:

1. **The Throw-It-Over-the-Fence Syndrome,** where the project is tossed from one functional area of responsibility to the next with no previous communication, commitment, or buy-in, on features, schedule, resources, or budget.
2. **The 50-to-1 Life Cycle Correlation,** where work is performed out of order, or is performed inadequately the first time around, so that it has to be repeated. The result is completed work often has to be undone and redone. The penalty for doing project work "out of order" is said to be $50 for every $1 it *would* have cost if done correctly. However, this is only an average; the premium, especially in product development, can be hundreds of thousands of dollars. Imagine the cost of a minivan recall to replace the little latch on the rear seat.
3. **Resource Bottlenecks,** where the project manager suddenly discovers that available resources are insufficient to meet project needs. I am constantly astounded by the number of organizations where:
   a. projects generate more than 90 percent of the revenues; and

    b.   almost all resources are shared and multiprojected, with individuals working on four or five separate projects in a week; yet

    c.   no effort is made to assemble and maintain a resource database (resource library) that reflects usage across time and that allows bottleneck identification and resolution. In such organizations, projects are daily being thrown into chaos because of the huge number of bottlenecks. The excuse is always: "What's the point in trying to plan my project schedule? When it gets to the drafting department (or contracts, or testing, or manufacturing, or programming, or documentation, or packaging, or ...), it's gonna get delayed anyhow until they have someone to put on it." The resource manager's only reasonable response to such a situation is to turn his department into a "black hole," from which not even information can escape. ("No, I can't tell you when your work will be ready, or who will be working on it. We'll call you when it's done.") It never seems to dawn on anyone that the lack of project planning and scheduling is precisely what ensures that there will be delays in each department.

4. **Resource Underutilization**, and resultant wasted funds, will occur repeatedly. Efficient use of resources in a project situation has less to do with how organized or hardworking those resources are, and much more to do with the project context. Resources must be utilized on the *right* project work and, in a multiproject environment, on the right project. Targeting the resources to the right work requires the employment of the critical path method (CPM) in scheduling, on each project and all projects in the organization's portfolio. If CPM is not used, resources will be expended on work of a less critical nature (from a scheduling viewpoint); in that sense, they will be underutilized.

5. **Abandoned Projects** will occur much more frequently. Make no mistake about it, the no. 1 reason that corporate projects are abandoned before completion is that planning was either totally inadequate or nonexistent. Frequently, I and other project management consultants are called upon to act as "project doctors"; we are hired to make a sick project well. Typically on such projects, no work breakdown structure (WBS) has ever been developed. As a result, no one knows what work remains to be done, *or even what work has already been completed*. Good consultants know exactly what to do; they go back to square one and, with the help of the subject matter experts on the project, develop a detailed WBS. But, what do senior managers do if they don't bring in a consultant? "Heck, this project's a turkey. No one can even tell me where we are on it. Spending any more on this

thing would just be throwing good money after bad. Time to pull the plug." And who's to say they're wrong?*

6. **Lack of Direction When Responding to Change**, which is by far the most important pitfall of inadequate planning. We are about to explore this idea in great depth, for herein lies the entire philosophical basis for project management. For the moment, it is enough to say that the project plan is like a roadmap. Its purpose is not only to show you the best route to where you want to go, but also, if you get lost, to show you the best way to get back on course. Without it, you are truly lost.

The benefits of project planning

Nobody likes to be wrong, and, if one's job is on the line, it is particularly important not to be wrong. Wrong planning, or estimates, for a project can be painful. The result can be overwork, underpay, blame, and unemployment. Yet project estimates are extremely dodgy; project requirements and information are constantly changing, making the best laid plans of mice and managers inaccurate. However, there's no use complaining about this; it's simply the nature of project work. If you can't deal with this, get out of the project business.

Unfortunately, this dodginess often makes project workers extremely reluctant to make estimates or predictions about their project. They are often afraid that, when things change, they will be blamed for the inaccuracies of their predictions, and their toes held to the fire.

To avoid that happening, project leaders often resist committing their expectations to paper. "Sure I've planned it," they say. "I've got it all in my head." But that doesn't allow them to manage a complex project, with requirements and schedules constantly changing, and other team members needing to understand just what *is* in their heads.

Yet, one must have sympathy for these project leaders. Because *they, in fact, are correct; they will be blamed, and their toes held to the fire if their predictions turn out to be wrong.* So they do what any reasonable person would do when placed in such an unreasonable position, they pad their estimates. Then, of course, Parkinson's law (work expands to fill time available) takes over, and two of the primary reasons for project planning, namely, shorter durations and more efficient resource usage, are torpedoed right up front.

---

* Actually, I am, to some extent. An article that I published in the September 1992 issue of *Project Management Journal*, "When the DIPP Dips," suggests that no project should be terminated without careful analysis of its DIPP (see Glossary). Without such analysis, project termination may lose more money than continuance.

Unfortunately, neither project leaders nor senior managers really understand the purpose of project planning. They are assuming that planning is performed in order to generate an accurate prediction of the future. Yet, on projects of any significant size and/or complexity, an accurate prediction of the future is just about impossible. *Things change.* While an accurate estimate is certainly preferred over an inaccurate one, from a project point of view, *an inaccurate prediction in a flexible format is one hell of a lot better than no prediction at all.*

This brings us to the heart of the philosophy behind project management, and it is crucial that this be understood. Projects can be large, complex, and expensive efforts. A failure can cause corporate bankruptcy, lost fortunes, and long waits in unemployment lines for thousands of people. Project management methods can greatly reduce the likelihood of such a disaster. However, *only* if the understanding of those asked to employ such methods is not such that it makes them sabotage the effort by such tactics as padded estimates.

The purpose of a project plan

What precisely do we want planning to do for us? What are the features of a good product plan that will generate the desired benefits? If we know that, we can work to improve our planning techniques.

Traditional project management consists of a "toolbox" of different techniques, discovered primarily by project managers over the years. These techniques were developed and their use spread precisely because they turned out to be helpful. So, with what characteristics of projects do project managers have to wrestle? There are two that seem particularly significant:

1. Typically, projects consist not of one type of work or discipline, but of many, all interrelated and interdependent, some dodgier and more nebulous than others.
2. All projects are subject to change, and managing those changes is one of the most difficult parts of a project manager's job.

These two realities of project managers' lives lead to two generalizations about the techniques they have developed:

1. They are likely to be universally useful, without regard to the nature of the project work involved.
2. They are likely to be of *particular* assistance in managing change, the greatest issue with which project managers constantly wrestle.

Having discovered many years ago that projects change, project managers have developed a methodology replete with techniques for managing these changes. How ironic, therefore, that the awareness of project uncertainty is often used as an excuse to avoid planning. "Heck," says the harried project leader, "this work is so nebulous and so cutting edge that who knows *what* we're gonna find once we start in. Whatever we plan now is bound to change, so there's no point in planning at all." This, I suggest, is roughly analogous to a medical school student saying, "Heck, there's no point in studying normal anatomy because all the people I see are going to be sick."

Now we need to ask, what is the project manager going to want to do when a change occurs?* What would be the main characteristics of a technique that would help her manage the unforeseen events that cause changes? The process that she will go through necessitates having a project plan in the first place. If she does, she will then follow seven steps. These seven steps provide us, quite serendipitously, with the happy acronym: A–I–M F–I–R–E. Seems like it might be a useful mnemonic.

A–I–M

When a project change occurs, the project manager, first and foremost, needs to be

- **Aware that such an event has occurred.** This can only be done by noting a "variance" between the plan and what has actually occurred. And some plans are much better at this than others. It is hugely important that this variance be identified as early as possible, while it can be remedied with the least drastic corrective action. To return to the medical analogy, the project plan works as a diagnostic tool, identifying the abnormality while it is still possible to treat the patient with the least invasive procedures. The earlier that the plan helps identify the variance, and the smaller the variance is when it's identified, the better the plan, and then the project manager needs to
- **Isolate the areas of impact.** You don't want to go around cutting tissue where there's nothing wrong. The philosophy behind management-by-exception is that it allows the manager's attention to be concentrated *only* where it's needed, and, in those areas, the project manager should now
- **Measure the variance.** This means the size of the difference between plan and actual in terms of changes in requirements, schedule slippage, or cost overruns, or, in Total Project Control

---

* When I ask this question in my seminars, I usually get an answer like: "Correct it!" This reminds me of the guy with the .44 Magnum again: Fire! Ready! Aim!

(TPC), in the expected monetary value of the deliverable (perhaps triggered by changed requirements or schedule slippage). All these changes should be analyzed and reported in terms of their impact on TPC's fundamental metric, the DIPP (Devaux's Index of Project Performance). This not only allows analysis on the basis of profitability, but also hitches the project's wagon to maximizing that metric.

Now, what kind of plan, what kind of qualities in it, will best assist the project manager in meeting her A–I–M? The answer: granularity, or detail, and quantification.

The broader and more general the plan, the greater the variance can grow before it becomes detectable. For example, a variance in a schedule that is measured in weeks and reported once a month may be as much as four weeks behind before it becomes evident. A two-week reporting period, on activities measured in days, is likely to turn up schedule slippage much earlier.*

Traditionally, quantification is in time units for schedule and in money units for budgetary cost, and variances are reported in these metrics. The TPC approach, by recognizing projects as investments and measuring all parameters in terms of their dollar impact on that investment, provides a single unit that quantifies the impact across all measurable variances, so that decisions can be made on the basis of comparative, profit-based data.

F–I–R–E

A variance having been identified, isolated, and measured, the project manager next needs to determine if action is needed, and, if it is, what is best. Again, the project plan is the crucial tool to assist in this.

Things the project manager needs to do include:

1. **Forecast the future impact** by analyzing what will be the effect of the current variance if no changes are made to the plan. Is the current variance a "one-time hit?" Will the one-week delay, or the $1,000 over-spending, remain at that level, or are these variances either the symptoms or the triggers of other problems? What if that one week delay means that the resources for a subsequent activity

---

* Scheduling and tracking on some types of projects is often much more granular. Nuclear power plant refueling outages, for example, are sometimes planned in activities that are measured in quarter hours, and reported on at the end of each eight-hour shift. This is due to the huge premium that is paid by the plant for being "offline," up to $1 million per day at some U.S. plants.

will no longer be available when they will be needed? What will all these variances add up to by project completion? And what will the impact be on the project DIPP?

2. **Investigate alternatives to the current plan** for routes to more satisfactory outcomes. Are they practical? What might the ripple effects be of employing such routes? Such analysis is done through "what-if" scenarios. Basically, these consist of introducing even more "variances" to the plan in order to see what beneficial effects there might be. Again, "beneficial" means in terms of the project's business value. If after all the alternatives have been investigated, the project manager is unable to avoid taking a huge hit on the project DIPP, this should call for immediate escalation to a project review board or senior management. If the project manager is able to maintain, or even raise, the original DIPP, is it only by diverting resources from another project? Perhaps from the critical path of another project, thus delaying it and perhaps lowering its DIPP? Should the project manager even have the capability to look at such an alternative? Who would approve such a change, and on what data? Both of these situations, therefore, should trigger:

3. **Review by senior management** if a threshold level built in on each project's DIPP is tripped. This review should occur any time that the project's expected profitability declines by more than a certain amount. However, senior management also should have its finger on the organizational profitability, as reflected by the multiproject portfolio DIPP. Any significant change (i.e., greater than 5 percent) in the DIPP of one or more projects should automatically be uploaded to the multiproject level and reflected in the portfolio DIPP. Senior management should have the ability to check this through a desktop information system. The profitability data, and delta, as shown in Table 2.1, should be available, and checked, each morning.

Any decrease in portfolio DIPP should be immediately traceable to the project change that triggered it. And, if senior management is uncomfortable with that change, an ad hoc review should take place. This may involve two or more project managers as well as resource and marketing managers. Notice again how it is the quantifiable nature of the project management data, especially the DIPP,

Table 2.1 TPC Senior Management Report on Portfolio Profitability

	Start EMV (000)	Budget (000)	Start EPP (000)	Cost ETC (000)	Current EMV (000)	Starting DIPP	Planned DIPP	Current DIPP	DPI	Current EPP (000)
Total Portfolio	$30,000	$17,000	$13,000	$9,200	$23,700	1.3	2.9	2.6	0.90	$6,700

that triggers the review by showing, in numerical terms, that (a) a change has occurred, and (b) the impact has been detrimental, and by how much. If necessary, this must lead to the project team to

4. **Execute a new plan** with whatever modifications (delayed deadline? Retargeting resources from another project? Hiring new or more expensive resources? Trimming features of the deliverable?) will result in the most satisfactory outcome, *not just for the one project, but for the organization's entire portfolio.* And the project manager must communicate the new plan to all affected parties (team members, suppliers, customer interface) who have not yet been notified.

What sort of plan will aid the project manager in performing the F–I–R–E procedures quickly, accurately, and comprehensively? The primary answer is *flexibility.* The more flexible the medium and format in which the plan resides, the easier it is to visualize future impacts, whether triggered by unplanned variances or input as part of the what-if analysis. Also, the faster and easier it is to document and publish the revision(s).

As far as the medium is concerned, the computer is obviously a wonderfully flexible tool. Project management software varies quite considerably in terms of its flexibility, but even the least user-friendly packages represent a giant advantage over the age of precomputerized project management, when schedules would be drawn in pencil on rolls of butcher paper, then taped up and down the corridors of the engineering department's building. Changes in those days (or, indeed, even during the early days of computers, when punch cards had to be used for every change) were a true nuisance, and what-if analysis all had to be conducted in the human brain.

Perhaps the very inflexibility of the premicroprocessor medium forced project managers to store their data in formats that had the utmost flexibility. The WBS and the critical path method network diagram (or PERT chart) are wonderful formats precisely because it is possible to see the impact of a change almost as soon as it is input to the plan. The WBS holds the scope and budgetary data in a way that allows for easy additions and subtractions, while the CPM network "flows" an initial change early in the schedule through to the end of the project, so that its impact can be measured.

Managing projects effectively puts a premium on managing the changes that impact every project. These can only be managed through the initial preparation of a project plan, which is detailed and in a flexible format and medium. In other words, you cannot A–I–M and F–I–R–E unless, first, you are ready to do so. That means knowing how to develop a good project plan in a suitable format.

Accuracy in planning is always better than inaccuracy. A completely accurate plan would obviate the need to ever manage change. However, the size and complexity of projects mean that all but the simplest projects are likely to undergo significant (if not major) changes during execution. A plan that was once thought to be accurate and, therefore was cast in concrete, becomes worthless with the first change.

A good project plan is one that provides the detail and flexibility to assist in the A–I–M and F–I–R–E process. It is a working document, to be repeatedly updated as the project goes along, and to be used as a tool for managing change.

Corollary and paradox

Now, let us turn to an oft-heard refrain from project planners:

> Some of this project is pretty straightforward, and I can give you a real solid plan for doing it. But some is nebulous, and so speculative that who knows *what* we will find. Since whatever we plan now is likely to change, there's no point in planning at all.

Suppose that our project planner is correct in diagnosing the situation: one portion of the project is simple and straightforward, the other is extremely dodgy. The $64,000 question is this: If the purpose of a detailed and flexible project plan is to help manage change, *where do you most need that plan?* On a project (or phase, or task) where the product scope is known, the work is straightforward, the resources committed, and the workers experienced? Or on a project where the work is nebulous, the technology cutting edge, and where we don't know what we are going to find, or where we don't control the availability of resources, or where whatever we plan now is bound to change? Surely it is obvious that, if a project is deemed so simple and straightforward that the plan is not likely to change, then we hardly have to plan it at all. It's the 99 percent of projects that *are* tough and nebulous that need a project plan as a working document to help manage the unpredictable changes that we predict are going to occur. And it is precisely those parts of a project that are most dodgy that need the most planning, the most attention, the most detail.

So we have a paradox: *The more likely it is that the plan will turn out to be inaccurate, the more benefit the project manager will get from a plan that is detailed and flexible.* The fuzzier the goals, the more important it is to plan. Again, the object of planning is not accuracy so much as a working document that can serve as a tool for dealing with the inaccuracies.

Empirical evidence for the value of planning when facing uncertainty

In June 2002, three years after the first edition of this book was published, the *Academy of Management Journal* published a paper based on research on 80 product development projects tracked over two years. The study, funded by the National Science Foundation and the Center for Innovation Management Studies, examined project management styles in dealing with uncertainties. Among the authors (Marianne Lewis et al.) conclusions were:

> Planned activities ... consistently enhanced the probability that a project would build technical knowledge, even under conditions of rising technical uncertainty. This finding supports Devaux's (1999) paradoxical claim that planning is most vital when a situation is difficult to plan. When a project's science and technology is nebulous, formal reviews and directive control may provide teams with vital direction.[1]

Storing the albeit uncertain planning data in the flexible formats of traditional project management tools, and then modifying the plan as triggered by changing circumstances, is a powerful way of dealing with the ambiguities and risks present in all complex projects.

How and what to plan

So much for the benefits of project planning, but what about the issue of *how* to plan? Here is where TPC alters the paradigm for even those organizations that currently utilize the full array of traditional project management planning techniques. The typical way in which a corporate project is undertaken is approximately as follows:

1. Someone gets an idea for a product, or a cost reduction program, or decides to respond to a request-for-proposal (RFP) from a potential customer.
2. Initial summary planning is conducted, often by the marketing department. This results in a business case that provides a thumbnail sketch of the "deliverable," estimates the potential revenue benefits, outlines the resources, and "guesstimates" the budget. The business case frequently also affixes a project duration, or delivery date.

3. The business case is then reviewed, usually by the customer or senior management. Sometimes, the important step of including those who will have primary responsibility for the development project (e.g., project manager, engineering department, manufacturing) is undertaken at this stage. However, all too often, that step does not occur until after final commitment to the project has been made. Then, those poor devils get saddled with the whole parcel, work scope, budget, and deadline, without ever being allowed to "buy out" of any commitments that seem unreasonable.

Once the project is "approved" (or the contract signed), a project team is "assigned" to it, with all its concrete-etched commitments. Why is a certain feature included? Why is June 15 the deadline? Why is the budget $3.4 million? Don't be silly. Because it is. Now go find a way to do it.

Such arbitrary project parameters result in two scenarios:

1. On rare occasions, the parameters might be too loose. In that case, Parkinson's law (work expands to fill time available) and its budgetary corollary will surely bubble out both the project plan and its implementation to its inefficient limits.
2. Much more often, the project will not fit within the preset parameters, and one, or more, or all, will "slip," whether formally or invisibly. Thus millions of dollars are lost every year through such "unmanaged inefficiencies."

To fully understand what is occurring here, it is necessary to return to the project triangle model we first discussed in the Introduction. Every project, no matter what the industry or work type, is a compromise among three variables: scope, time, and cost, as shown in Figure 2.1.

Again, the two scopes (product and project) represent the features of the final product or result and the total amount of work to generate it. Cost is the "budget" measuring total resource usage, and time is the total elapsed time, from initial project implementation to completion.

Figure 2.1 The project triangle.

Traditional project management deals only with the two sloped sides. Additionally, although the term *cost/schedule integration* is a common term in traditional project management, the "two-sided" approach offers no such benefit.

Resource usage can be turned into cost and quantified in dollars (or euros or yen) so as to provide a single unit across all resources. This lets us compare the expense of nuclear physicist work hours with janitor work hours, and also pounds of nails and interest on loans with janitor work hours. However, time is estimated and measured in hours, days, weeks, and months. In order to truly integrate cost and schedule, one would need to be able to use the same units for both so as to make decisions involving trade-offs between them. However, the two parameters use completely different metrics: Is spending $20,000 to save a week of time worthwhile?

Unless we know the value of time on the project, how can we analyze such a decision? Traditional metrics are no more capable of comparing cost with schedule than of apples with orangutans.

The only way to truly integrate all the parameters of a project is to deal with the three sides of the model in a single unit. However, in traditional project management, scope is not quantified at all. It is usually documented as a list of the "deliverables" (in U.S. government-related work, sometimes called a C-list or contract list, consisting of CLINs, or contract line item numbers). How do you quantify all the different forms that work scope on a project can take? What single metric could be used to unite depth charges with propeller blades, with missile telemetry software, with coats of paint, with bulkhead penetration test results, with toilet seats? This is what a true method of quantifying scope needs to provide. Further, one should then be able to take these distinct forms of deliverable and work scope, mix them all in with duration and budget, and come up with a single measurement for each aspect of a project, allowing a ratio for comparison, contrast, and trade-off with all other aspects of the project. This is what TPC offers.

Scope/cost/schedule integration

Scope, both product and project, is the foundation on which the whole project rests. It is the reason we do the project—because we want to obtain the value that will accrue from the product in the form of an additional "thing" that generates revenue or savings, or workforce experience from doing the work. In the triangle model, it is the baseline of the triangle. *That is exactly what the scope is; the basis for doing the project, the value its completion is expected to add to the organization.*

In a company whose revenue is completely project-generated (as is the case with most product development companies, as well as many other types), the total organizational bottom line, its profit, is the sum of the value produced by all the individual projects minus the operating costs.

Once we recognize this, two things come into clearer focus:

1. Quantifying scope is *important*. It is directly related to profitability. In a project-driven company, if you haven't quantified project scope, you cannot accurately estimate, or work to increase, profit.
2. The metric used to quantify scope is *the dollar*. To be precise, the *expected* dollar that measures the value that the project is undertaken to generate.

Now, how one goes about estimating the value is different on every project; a topic of its own beyond the scope of this book. Obviously, it often relies on experience, research, and guesswork. But this information is the driver of the entire project, and, obviously, some estimating of expected value is implicit for any project. Without an estimate that the final product will be worth more than the cost of the work we have to do, no project could ever be budgeted accurately, and, in a world of limited resources, no decision between two competing projects could ever be made without assumptions about their comparative expected values.

"But," comes the cry, "such estimates are likely to be inaccurate."

So try to make them accurate. If the budget for a project is likely to be $5 million, surely it's worth an additional 1 percent, or $50,000, to discover if the expected value is $40 million, $14 million, or $4 million. Then perhaps we can discover how to add a few million by changing the project's scope or schedule.

Besides, the project's budget and duration are only estimates, too. Every project budget includes indirect costs, such as specific overhead rates (50 percent of salary? 100 percent), which are obviously the crudest of estimates. Does anyone really think that every employee, from plant manager to janitor, at the New Jersey plant always has an overhead cost that is exactly 50 percent of their salary? But we have gotten used to accepting overheads estimates because they are useful.

The same is true of value estimates; estimating them has value. When a good poker player decides on whether to call, raise, or fold, he doesn't *know* exactly what is going to happen, or what the size of the pot will be when it's won, but he certainly uses all the available data, including the current size of the pot (the global market), the cards showing (current competitive products), and the proclivities of the other players (demand volatility) to estimate the growth potential of the pot and to determine his actions. Poker is reputed to be a game of luck, yet the player who performs this analysis best almost invariably walks out a winner at the end of the evening. Surely a multimillion-dollar project is worth more analytical effort than a dollar-ante poker game.

We shall resume the subject of expected monetary value (EMV) analysis when we talk about TPC's value breakdown structure (VBS) in the next chapter. But, for the moment, we can see that our quantification of scope by EMV gives us metrics for two sides of the triangle, scope and cost, in a single unit: money. Also, expected monetary value minus cost equals anticipated profit. We can analyze scope-adding or cost-cutting measures according to their impact on these two sides, and thereby increase or decrease the anticipated profit. Also, we can compare two potential projects according to their anticipated profits and elect to pursue the more profitable one.

What about the third side of the triangle: time? The answer is that the expected value of a project is based not only on the work scope, but also on *when* that work scope is completed. Thus, the estimate of expected value is time dependent: Deliver your Christmas toy to the stores by November 20 and you will generate revenues of $20 million; delay delivery until December 20 and you might make $2 million. Almost every project will vary in expected monetary value depending on completion date, even if only because it's that much later before the benefits of the work (e.g., cost savings or decreased personal labor) begin to accrue.

On some projects, delay will decrease the EMV at an even rate. Conversely, sometimes the cost of one day of delay can be precipitous. If a space probe is to be launched to visit a passing comet on April 15, then the probe's EMV is constant through that date. However, it may drop to zero if the launch date is missed and the comet is no longer reachable. In such cases, the only value in compressing the probe's schedule to earlier than April 15 is the reduction of the risk of being late (which does have value). If the project slips by so much as one day, it's worth the entire EMV to move the completion date back to the 15th. How to accomplish this, and what decisions can be justified in such circumstances, will be covered in a later chapter.

All the data regarding *each* project's EMV, and *every* project's EMV, across the entire portfolio, should be assembled up front as part of the TPC business case, and then should be tracked and, if necessary, updated at regular intervals throughout the project. It doesn't make much sense to have regular and up-to-date project information, but to allow the market information that is driving the project to be many months old. It is senior management's responsibility to make sure that the following data are collected, and then used, for every project as soon as it is approved for funding:

1. What is the expected monetary value of the project?
2. As of what date?
3. What is the plus or minus value for each unit of delay or acceleration?

If the EMV and delay/acceleration cost/value change, project analysis needs to reflect that change. The project manager should immediately be made aware of such changes, and possible modifications to the project plan analyzed.

Planning and tracking the DIPP

The index for assessing scope/cost/schedule integration for each project is the Tracking DIPP:

DIPP = Expected monetary value (as of current completion date)/ estimate-to-complete (ETC)

At the start of the project, the estimate-to-complete is the project budget. As implementation proceeds and work gets done, the estimate-to-complete should (hopefully) decrease, and the accrued costs become "sunk" costs, unrecoverable expenditures that, being unalterable, should only impact on the analysis of future actions if they assist trend analysis.

The future profitability of the project is the EMV *minus* the ETC, but the DIPP is an index of the efficiency with which the project resources will be utilized until completion. A DIPP of less than 1.0 means that the project will cost more to complete than its expected monetary value. Such a project should either be changed to increase the EMV by adding valuable work scope items, cut costs by reducing resource use, decrease delay by adding resources—or terminated.

By analyzing the EMV, ETC, and delay/acceleration cost/value, the project manager should always endeavor to "invest" the resources at her disposal in the best possible way to maximize the DIPP. Regular project reporting on the DIPP, period by period, should be standard operating procedure.

Additionally, the initial plan should include a forecast of what the DIPP is anticipated to be at each reporting point until the end of the project. (It, of course, should increase as costs are "sunk" and the ETC decreases, thus reducing the denominator.) A histogram, such as the one in Figure 2.2, should be produced with each report, showing how the DIPP is tracking against both the original and the current plans.

TPC at the organizational level

The same sort of analysis that is performed at the project level also should be done at the macro-, or multiproject portfolio level.

The portfolio DIPP indicates the expected profitability across all the projects. Senior management should be expected to analyze how current resources might be better "spread" to take advantage of the diverse project delay/acceleration cost/value factors. Senior management then can

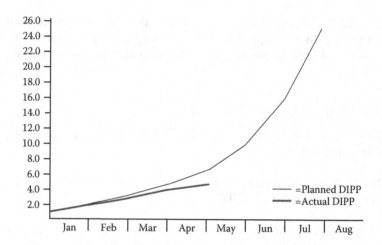

Figure 2.2 Histogram showing actual DIPP versus planned DIPP over time.

decide to delay (or terminate) some projects and accelerate others so as to target the resources to the best projects for the purpose of maximizing the portfolio DIPP.

The DIPP is *not* identical to organizational profitability. Given fixed costs, the DIPP indicates how much profit those costs are expected to generate. However, senior management also has the option of increasing costs by investing in additional resources. In such an event, the organizational DIPP (EMV divided by Cost ETC) might actually go *down* even as raw profitability (EMV *minus* Cost ETC) increases.

The following chapters will explore the specific techniques of TPC planning, but the two overriding concepts that we must bear in mind include:

1. The entire organization engaged in project work should at all times be driven by optimization of the multiproject portfolio's profitability.
2. Project profitability is a compromise among work scope, time, and cost. If work scope is quantified as expected monetary value, the project can be measured and optimized by using the DIPP.

Reference

1. Lewis, M., A. Welsh, G. E. Dehler, and S. G. Green. 2002. Product development tensions: Exploring contrasting styles of project management. *Academy of Management Journal* June 1: 25.

chapter three

Overview of planning the work

I mentioned earlier that every project is a compromise among the three variables of scope, time, and cost and may be pictured as the triangle in Figure 3.1. I mention it again not because I want to bore you to tears, but because this paradigm drives *so much* about the project context, whether within one project or throughout a project-oriented organization, that it keeps resurfacing as the jumping-off point for further issues and discussion.

Take the project as a whole, or any one of the myriad and diverse work tasks within it: how long will it take and how much will it cost? How does one go about getting even a "ballpark" estimate?

Every project is unique, consisting of a unique work scope performed under unique circumstances. Therefore, estimates about either time or cost also will be unique, dependent on the project-specific scope.

"Got some work for you to do. How long's it gonna take, and how much is it gonna cost?" Well, that depends on what the work is, and the project leader who provides estimates without an adequate definition of scope is signing his own pink slip.

Thus, the first task of the project planners is to define the scope, both product and project, in detail. The greater the detail, the more accurate the estimates that can be generated. But, make no mistake, *defining the scope is the hardest part of project planning.*

Let us now assume that our scope-defining effort has been successful. What we now have is the total product scope divided into its several components and subcomponents as shown in Figure 3.2.

Having identified the components of the product scope, the next step is to determine precisely what work, or project scope, must be performed to generate the components and ensure that they are meeting the required specs. This is accomplished through a product-based work breakdown structure (WBS). First the major product components are decomposed into their subcomponents. The group of lowest-level subcomponents on each branch is called a *work package*, a term that some find a bit confusing because it actually identifies no work. It is, however, a package of work, a product at the lowest decomposable level. All further decomposition of a work package is work, the project scope activities that must be performed to produce it. As Figure 3.3 shows, the product-based WBS reflects this process of going from product scope to project scope.

Figure 3.1 The project triangle.

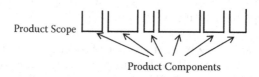

Figure 3.2 The product scope base.

There is some disagreement in project management circles as to whether the definition of a WBS is limited to the product scope (i.e., the work package level and above) or if it also includes the project scope. Because each item of project scope is explicitly performed in order to generate a work package, it seems clear to me that a WBS extends beyond the work package level and down to each individual work activity. But as long as it is understood that every activity is a decomposition of a work package, whether or not the work is part of the work breakdown structure is a matter of definition. (If it's not, why do we call it a *work* breakdown structure? Perhaps a product breakdown structure without the W would be a better term?)

With the project scope decomposed into detailed work activities, the planning team should now be in a position to generate time and cost estimates for each activity. The time estimates are of the elapsed time of the activity's duration, and the cost estimates are of the dollar value of the resources required to complete the activity's work scope. In effect, each

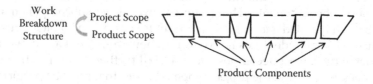

Figure 3.3 The product scope decomposed into the project scope through the WBS.

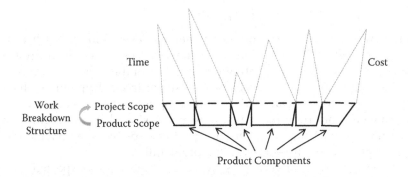

Figure 3.4 Building the project triangle from each component.

activity becomes its own project, with its own little triangle, as shown in Figure 3.4.

Each of these triangles, with their time and cost estimates, is now integrated into a total project plan. This will ultimately be done by scheduling all the work through the critical path method (CPM) and computing the budget through activity-based resource assignments (ABRA) and activity-based costing (ABC).

Let us suppose that we are a planning team working for Megaprodux, Inc. We have been assigned to develop a new product (MegaMan) and, after many days of effort, we have developed a plan that includes a schedule of 35 weeks with a budget of $3.5 million. Let us further suppose that this amount of time is deemed unacceptable, by the customer, by senior management, or due to some fixed market window, such as a space shuttle launch or the Christmas shopping season. We must reduce our schedule by, say, six weeks. Utilizing critical path method to its fullest functionality, we "crash the critical path" with additional and more expensive resources until we have a 30-week schedule, but now with a budget of $3.8 million. And when we produce our new plan for approval, it is vetoed. We are told that our mandate is to perform the project in 30 weeks for the original budgeted cost of $3.5 million.

If we have *really* done a thorough job of utilizing the critical path method for scheduling, and done all the "fast tracking" (simultaneous activities) we can, this leaves us but one choice: the third side of the triangle, or scope. Some of the components, or subcomponents, or features, or quantity, or quality testing will have to be trimmed from the planned scope in order to meet the mandated parameters of time and cost. However, where do we conduct such pruning? And what will its impact be? And does it make sense that we have been given this mandate in the first place? Before we answer these questions, let us see how Total Project Control (TPC) addresses this problem.

Quantifying the project triangle

This is where traditional project management, even with cost/schedule integration, offers little help, but where the *scope*/cost/schedule integrated plan of TPC allows for a quantified comparison of the trade-offs involved in different solutions to the problem. If, as suggested in the previous chapter, the expected monetary value of the project has been calculated and incorporated into the project planning process, we have a baseline for determining what impact other changes to the project may have on its most important feature—its expected profitability.

Expected profit, of course, is expected value minus expected cost. And in the project triangle, *COST* (which is really resource usage) is nicely quantified into dollars or some other monetary unit. If we have the expected monetary value of the project also monetized, then:

- Two of the three sides can be analyzed using the same unit.
- One side minus the other will give us the expected project profit.
- One side divided by the other (EMV divided by Cost ETC) will give us our profitability index, the DIPP (Devaux's Index of Project Performance).

So, now we have two sides of the project triangle quantified in the same all-important unit, as shown in Figure 3.5. But what about the third side to the triangle? How should *TIME* be quantified on a project? Well, time always *is* quantified—as weeks, days, hours, minutes (Figure 3.6). However, that does not really help us; how do we translate such units into dollars?

It was Benjamin Franklin, more than 200 years ago, who gave us the answer: "Time," he said, "is money." How, exactly, is *TIME* money on a project? As we suggested earlier, by delaying the point at which we start receiving the benefits of the completed scope.

The answer leads us to a crucial feature that distinguishes the TPC business case for a project from the traditional one. Most business cases for projects (and, of course, we must recognize that many corporate projects are embarked upon without ever having *any* kind of business case)

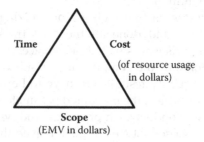

Figure 3.5 Quantifying two sides of the project triangle.

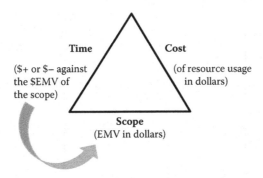

Figure 3.6 Quantifying all three sides of the project triangle.

will have a paragraph or column discussing the benefits of the project, and will often quantify these into an expected monetary value. It is crucial, however, to recognize that the expected monetary value of the project is *not* a constant; it is a variable, dependent upon exact details of the scope and the delivery date of the final product, and that delivery date is determined by the total elapsed time of the project. Therefore, the *TIME* side of the project triangle must be quantified, unit by unit, per its plus or minus dollar effect on the expected monetary value (EMV) of the project. That quantification belongs right up front as part of the business case, where it can be used by the project manager and planning team to generate an initial plan that is as valuable as possible (Figure 3.7).

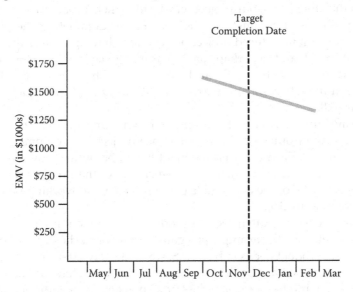

Figure 3.7 Delay Curve 1: EMV variance on a fixed price contract.

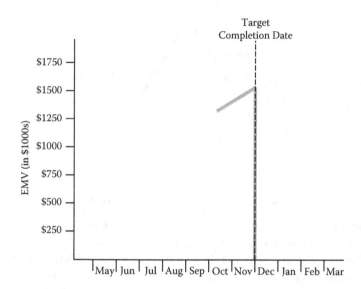

Figure 3.8 Delay Curve 2: EMV variance for a project with an acceleration cost.

Just how this plus or minus impacts the EMV varies from project to project. Delay Curve 1 represents a contracted delivery for a specific customer. The impact of a two-week delay might merely be the time value of money reduction for the same dollar amount received two weeks later. Earlier completion of such a project would result in a similarly small "acceleration premium."

On the other hand, our project may be to launch a space probe to take close-up photos of a comet passing near Earth. Completing the project early may result in reduced project profit, due to having to store the satellite until launch and to perhaps having its energy cells erode. However, a probe that misses its blast-off date by one day will have its EMV reduced from $20 million to zero. This we can consider as Delay Curve 2, as shown in Figure 3.8.

Delay Curve 3 occurs, for example, in refueling projects at nuclear power plants, or other similar maintenance projects on equipment, such as oil refineries. Each day that the plant has to be offline may represent the loss of $2 million. In such a case, *every* day of the project's duration reduces the EMV of the deliverable (i.e., a refueled and working plant) by $2 million (Figure 3.9).

Delay Curve 4 is often seen in product development projects for the retail market. Take the example of a game or toy for the holiday shopping season. The product needs to be in U.S. stores the day after Thanksgiving Day, when the bedlam begins. Our sales department tells us that to be one week late implies a revenue loss of 20 percent. Each additional week late equates to a similar loss until, after five weeks, the shopping season

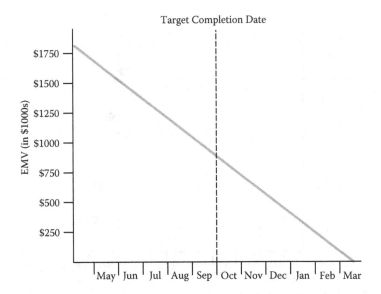

Figure 3.9 Delay Curve 3: EMV reduction for unavailability of revenue-generating equipment.

will have been missed and revenues reduced to zero. In addition, there is a small reward to be gained through early delivery; we can estimate that each week that our product is in the stores *before* Thanksgiving will increase our revenues by 4 percent. With an EMV of $20 million, our delay cost is $4 million for each week after Thanksgiving, with an acceleration premium of $800,000 for each week before that date. This type of delay curve is pictured in Figure 3.10.

Delay Curve 5 is the situation fairly typical of pharmaceutical and other R&D-based product development projects, where the first product to market will enjoy a huge boost to revenues, the second to market a smaller share, and so on. The pharmaceutical company developing the product knows that competitors are at work in the same field, but does not know how close to market they are. In such a situation, per unit delay cost must be estimated on the basis of probability and risk:

- What can we expect revenues to be if we are first to market versus second to market?
- What are the probable dates on which our competitor might deliver an identical product to market, and how does that probability change per unit of time?

Such information should be the duty of our marketing department.

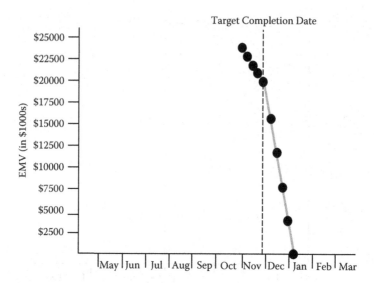

Figure 3.10 Delay Curve 4: EMV variance for a project development project with a seasonal market window.

- It has researched the situation and has estimated that, if our product is first to market, it will result in an expected monetary value of $100 million. Being second to market will lower the EMV to $40 million.
- It has also determined that there is a 10-percent chance of our competitor reaching market May 1, a further 50-percent by June 1, a further 25-percent chance by the beginning of July, and 5-percent more for each of August, September, and October. Figure 3.11 shows that, based on these numbers, our project's EMV will be $100 million if delivered by the end of April, $94 million in May, $64 million in June, $49 million in July, $46 million in August, $43 million in September, and $40 million in October.

Based on these data, Figure 3.12 shows the delay cost curve for each month from April through October. The first month is worth $6 million, the second an additional $30 million, the third $15 million, and so on.

Most projects fit into one or another of the five delay curve profiles shown. If we know our delay curve, then, depending on what delivery date our schedule is currently headed for, we can calculate the maximum amount that we should spend on additional resources to accelerate our schedule. In my experience, this amount is almost always much more than the resources would cost (and almost always more than the organization that has *not* had the situation spelled out for them, in TPC terms, is willing to allocate).

	In APR	In MAY	In JUN	In JUL	In AUG	In SEP	After OCT 1	
First to Market	$100M 100%	$100M 90%	$100M 40%	$100M 15%	$100M 10%	$100M 5%	$100M 0%	
	(=$100M)	(=$90M)	(=$40M)	(=$15M)	(=$10M)	(=$5M)	(=$0M)	
		EMV = $100M	EMV = $94M	EMV = $64M	EMV = $49M	EMV = $46M	EMV = $43M	EMV = $40M
	(=$0M)	(=$4M)	(=$24M)	(=$34M)	(=$36M)	(=$38M)	(=$40M)	
Second to Market	0% $40M	10% $40M	60% $40M	85% $40M	90% $40M	95% $40M	100% $40M	

Figure 3.11 Diagram computing EMV variance based on value and probability of being first or second to market.

In addition to spending more money on resources, there is another way of shortening the schedule: pruning scope. Increasing resources tends to reduce profit by increasing cost. Pruning scope also tends to reduce profit, but by reducing EMV, a product with fewer features, or less reliability, or reduced advertising is likely to generate less revenue. Which

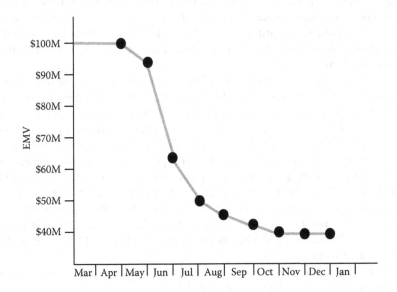

Figure 3.12 Delay Curve 5: EMV variance based on value and probability of being first or second to market.

should a project manager elect to do, add resources or cut scope? Answer: Whatever leads to a larger expected project profit and/or DIPP.

Estimating the minimum value/cost of time

The project manager and project team do not necessarily need to know the expected monetary value of the project, though that is definitely information that the sponsor/customer and anyone else helping to fund the project should have estimated. However, the project manager and team definitely need to have an estimate as to the value/cost of time for completion earlier or later than the target date. Indeed, these are the trenches in which the project operates, within the parameters of scope, schedule, cost, and risk.

The value/cost of time should be in the project documentation—either the project charter, the project contract, or the project business case. It is an essential element of the TPC business case. Unfortunately, most sponsors/customers do not understand that they need to provide the project manager and team with this estimate and that omitting to do so leaves the team unable even to look for the opportunities that might enhance the project's value.

If the sponsor/customer, however, has not specified the value/cost of time, the project manager still needs that information. It is crucial for justifying resources and making other trade-offs among the parameters of the project. Under such circumstances, my recommendation is that the project manager should assemble some "back-of-the-envelope" estimates regarding the minimum possible value/cost of time and formalizing those calculations in a memo to the sponsor/customer for approval or modification.

The project manager may not be able to accurately estimate the importance of time to the sponsor, but she should be able to estimate a minimum value if she has been given two data items that are fairly standard in the documentation of any project: a project duration or deadline and a budget.

As an example, let us assume that the project is one that has been executed at many organizations over the past two decades: implementation of a new enterprise resource planning (ERP) system such as SAP or Oracle's® PeopleSoft across our entire medium-sized organization. Typically on such projects, no one tells the project manager and team anything about either the project's EMV or the value/cost of time, but let us assume that the mandated parameters are a duration of 12 months and a budget of $5 million.

To estimate the value/cost of time, we first need to have an estimate of the expected monetary value for the entire project if completed in 12 months. If the budget is $5 million, we know that the EMV *has* to be greater than $5 million because every project is an investment; no one ever knowingly funds a project that is expected to produce less value than it costs.

But how much more than $5 million will this project be worth? Most capital projects are justified on the basis of what is called a *payback period*. This is the amount of time that it is estimated it will take for the value produced to equal the cost. Payback periods vary according to the length of time it will take before the new equipment will be obsolete. For a blast furnace, it might be 40 years. These days, for any type of computer system, far more common is three years or, at a maximum, five years.

If we know that the standard payback in our organization for enterprise-wide software systems is three years, we should assume three years for our ERP project. However, if we don't know, we should take the more conservative five years. That means we will assume that the value that the new system is intended to generate must be, at a minimum, $5 million over the 60 months after implementation.

Of course, we could simply take that $5 million and buy U.S. Treasury securities and be assured of generating more than $5 million. So, the truth is that this project must be expected to generate more than $5 million of value over 60 months. How much more? Well, 4 percent per year seems like a fairly minimal return to assume.

4% × 5 years = 20%
20% × $5 million = $1 million
$5 million + $1 million = $6 million
$6 million ÷ 60 months = $100,000 per month, or $23,076 per week

If the project takes 13 months instead of 12, the new system will reach the same point of obsolescence after 59 months of use instead of 60 and, therefore, the value it generates may be expected to be reduced by $100,000. That is its minimum delay cost.

However, if the project takes only 11 months instead of 12, the new system will not reach the same point of obsolescence until after 61 months of use instead of 60 and, therefore, the value it generates may be expected to be increased by approximately $100,000. That is its minimum acceleration premium.

Based on these estimates, a decision to spend an extra $10,000 on resources that will shorten the project duration by two weeks would generate an extra $46,152 of expected monetary value and an extra $36,152 of expected project profit.

> Dear Sir, on the basis of my back-of-the-envelope calculations, it seems to me that the minimum amount that any week of acceleration or delay on this project would be worth is approximately $23,000. If this

makes sense to you, that is the number that I will use in order to seek opportunities to improve the value of this project through scheduling trade-offs. If there is a different estimate of the value/cost of time on this project, please let me know so that I can use your estimate for seeking such opportunities. Under any circumstances, please be assured that any suggested changes to budget or schedule will always be forwarded for your approval before they are implemented.

And, frequently, of course, the project sponsor will respond by saying that the value of time to *him* is much greater than $23,077 per week. And the project team will then use his numbers to seek opportunities to enhance both the project's and the project team's value.

Optimizing the DIPP at the micro level

Let us now return to the problem we left earlier, where we, as a planning team for Megaprodux, Inc., were being ordered to cut our 35-week MegaMan project down to 30 weeks while maintaining our budget at $3.5 million. Let us now also incorporate into this scenario the information that Megaprodux is a toy company, that Week 31 is the start of the holiday shopping season, and that our deliverable, MegaMan, is expected to generate sales of $10 million. We will assume that Delay Curve 4 (Figure 3.10) is applicable. All this information belongs in the TPC business case, and can be succinctly summarized in the fashion displayed in Figure 3.13.

First of all, if we are to do as ordered and complete the project in 30 weeks for $3.5 million, we could cut scope. However, we want to cut it in the way that reduces the project's value as little as possible. Now, how do we determine which of the little activity triangles from our WBS is adding

---

MegaMan Development Project

EMV: $10,000,000 if completed at end of Week 30

Delay penalty: 20%, or $2,000,000 per week
Acceleration premium: 4%, or $400,000 per week

Target budget: $3,500,000

Figure 3.13 Summary of the TPC business case for the MegaMan development project.

the appropriate amounts to time and cost such that their removal will give us the schedule and budget we need while reducing the value the least?

When we "drill" down to the micro, or activity, level of the project, the relationships between components and between the work activities required to produce them are closely intertwined, both in terms of expected monetary value and schedule. The impact that removing a component or activity has on a project depends not only on the component or activity itself, but also on the rest of the product and project.

What, for example, is the value of a staircase in a house? It depends very much on where the staircase leads. If the house is a one-floor ranch, and the staircase leads nowhere, its value is, at most, decorative. On the other hand, if most of the important rooms in the house are on the second floor, then the staircase acquires a value almost equal to the total value of the second floor (unless there is also an elevator). The issue is: How much would the house be worth *without* the staircase, if it had all the other rooms and features, but no staircase? How much value is the staircase adding to the house? This is called its *value-added*. We will explore how to compute this value-added in greater detail shortly.

What about the *TIME* side of the activity's triangle? How interconnected is that? Again, whereas a project's time is its duration, an activity's duration may have no significance whatever when attempting to shorten a project's schedule. It all depends on *where* in the project schedule the activity occurs. As we shall see shortly, if the project is scheduled through the critical path method, only critical path activities impact the project schedule. How much they impact it depends on far more than simply the activity's duration. This impact is one of the key data items of the TPC methodology, and is called the activity's *critical path drag.*

We will cover drag in great detail in both the chapters on scheduling the work (Chapter 6) and scheduling the resources (Chapter 9). For the moment, let us just say that drag is the amount of time an activity is adding to the duration of the total project, or, conversely, the amount of time that could be saved by *removing* an activity from the project schedule.

So far, however, we have only addressed the time of an activity, not the *TIME* of an activity. You will recall that when we wanted to analyze the project in a scope/cost/schedule manner, we had to deal with time *not* as a duration (in weeks, days, hours, etc.), but as an impact on the EMV of the project. So, too, the *TIME* of the activity triangle should be measured not simply in drag units, but in another TPC metric: drag cost (*the amount of money by which the project's value is being reduced as a result of having to perform this activity, with X number of units of drag*).

This is hugely important, because the cost of the time to do a project is usually far larger than the cost of the resources (i.e., COST) to do the project, and that correlation extends down to the activity level in even more dramatic numbers. A critical path activity may be utilizing $30,000 in resources over its three weeks of duration. However, if two of those three weeks represent time added to the project schedule, it is not exaggerating to say that, in a product development project, each week could represent $500,000 in irrecoverable revenues, or a total activity drag cost of $1 million for the two weeks of drag. As Ben said, time is money.

Activities cannot, then, always be managed to maximum profit. Well, they *can*, but it depends on how you define profit. What is the profit on a staircase within a house? What is the profit on the left wing of an airplane? Without the project, the activity may be worth nothing (or very little). The activity gets its value by adding value to the rest of the project, but that value is offset by:

- the cost of resources required to perform that activity and
- the reduction in the project's EMV due to the amount of time it is delayed by the performance of that activity.

This leads us to the activity's *equivalent* of profit: its net value-added (NVA). The NVA of an activity is its value-added minus the sum of its resource cost and its drag cost.

NVA = value-added – (cost + drag cost)

The NVA of an activity may change as the project is implemented. An activity that has a value-added of $200,000, a budget of $20,000, and two weeks of drag may start with an NVA of $180,000 if the initial schedule will meet the deadline and there is no acceleration premium. However, if there is a delay cost of $100,000 per week, and the project's critical path slips two weeks, suddenly the NVA will be:

$200,000 – ($20,000 + $200,000) = – $20,000

If an activity is on the critical path, its *true cost* is the sum of its resource costs plus its drag cost. This activity has a true cost that is greater than its value-added. Either this activity should be removed from the project or another change should take place in how it is scheduled. However, this all requires careful and detailed oversight of the project. Also, most traditional project planning and project management software does not support such data items as EMV, the DIPP, activity value-added, delay cost, acceleration premium, NVA, drag cost, or even simple critical path drag calculation. Ideal, of course, would be a software package that would not only handle such input and output, but also would send an alert whenever any component or activity declined to a negative value-added (or a value-added below a level preset by the user). Without such software,

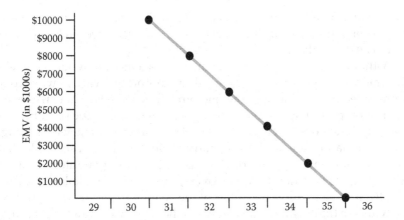

Figure 3.14 Chart of MegaMan project EMV, weeks 30–35, based on Delay Curve 4.

much project effort on slipping projects is likely to be wasted on work of negative value.

Let us take the MegaMan project as an example. With a $10 million EMV at 30 weeks duration, and loss of 20 percent for each week longer (from Delay Curve 4, Figure 3.10), we can see from the numbers in Figure 3.14 that senior management is wise to insist on a maximum duration of 30 weeks.

However, limiting the budget might not be the wisest idea. The tactic of limiting costs while requiring shorter durations is only successful when projects have been badly planned. It assumes padded duration estimates and poor application of CPM techniques. However, if senior management really assumes that it can shorten the project by five weeks simply by reducing the inefficiencies of the project manager, it should just fire the project manager and start over.

(Of course, senior management is often *right* in such an assumption, but it must shoulder the blame for project management knowledge and techniques not being standard operating procedure within the organization. Standardized project management procedures, including reporting and oversight, a good project management software package, and training for both its project teams and *itself* in the intricacies of project management would go a long way toward making projects shorter, cheaper, and more profitable.)

If on our MegaMan project the planning team has applied the techniques of traditional project management properly, then requesting the additional $300,000 in order to shorten the project by five weeks is probably both reasonable and wise. Even at the gut level, without TPC metrics, it doesn't take a genius to decide that this project needs to be done by the start of Week 31, and if the additional $0.3 million needed to trim the five weeks is enough to drastically reduce the profit, then we probably don't want to be doing this project in the first place.

Without realizing it, though, senior management may have made a much more costly decision. Despite all the trouble and care to which our planning team has gone, we now find ourselves being told to make this a $3.5 million project, and the only way to do that is to reduce cost by cutting scope. Out goes MegaMan's elaborate costume and fancy packaging, along with half of the advertising budget. Now we have a schedule that will have our toy in the stores on time, as well as a toy guaranteed to be on the shelves, marked down by 90 percent, weeks after the holiday season ends. Even if the retailers take most of the beating this year, Megaprodux, Inc., will take the beating *next* year, when the retailers shy away from such a loser.

Most of the time, this sort of thing happens without anyone in senior management *even being aware of it.*

Again, scope is the ignored stepchild of the traditional project management approach. It is regarded as a constant, which allows managers at all levels to tinker with schedule and budget while pretending that they are leaving scope unchanged. In the MegaMan example, the changes are pretty drastic and should be extremely visible. However, if senior management does not bother to look, the results can be disastrous.

The pruning of scope is often subtle and insidious. Design is rushed, testing is shortened, corrections are not double checked, and quality is thoroughly compromised, all without leaving telltale evidence until the product collapses on the shelves, or while little Jenny is playing with it. Which might be fine, except that Jenny's mom takes it back to the store, the store reports it defective, and removes it from the shelves.

Using TPC on the MegaMan project

If Megaprodux, Inc., mandated the use of the TPC approach, the entire MegaMan scenario would be different.

In the first place, our planning team would never have submitted a project plan with a 35-week schedule. The TPC business case from which we were working would have guided us by quantifying the lost revenue for each week it was late. With the DIPP as our guiding star, we would have planned toward maximized profitability, and very likely generated a schedule of 30 weeks, with minimum cost. However, even that might have been insufficient. Remember, in addition to the delay cost for the weeks beyond Week 30, Delay Curve 4 offers an acceleration premium of 4 percent per week for each week *less* than Week 30. The entire EMV picture is shown in Figure 3.15.

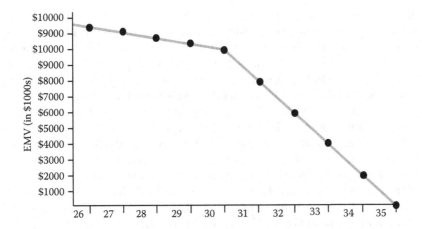

Figure 3.15 Acceleration/delay curve weeks 26–35 based on 4-percent gain and 20-percent loss of EMV per week.

Now, with an additional $400,000 per week to be made through schedule acceleration, it may be possible to augment MegaMan's profit even more. Sure, it's going to be tough to squeeze additional weeks out of an already tight schedule, but with *up to* $400,000 per week in additional resources to be targeted to just the *right* activities, something might be possible. Let's see, spend an extra $350,000 each week for five weeks and make an additional quarter million. Hmm. Know any widgeters who might be willing to work for $350,000 per week?

The other thing that the TPC approach would do is make it clear that cutting the costume and packaging, and halving the advertising budget might not be smart moves. The value-addeds of those activities may slice deeply into the EMV, which would be immediately apparent through a reduced DIPP.

Of course, if no one is *watching* the DIPP, any decision is likely to be wrong.

There may be components and activities that could be eliminated without serious penalty. Such candidates would be the ones that the TPC approach would point out as having low NVAs. Again, a software package that lists NVAs in ascending order would be most helpful, but first, such data have to be input to the project plan.

Conclusion

The reader should by this time have a pretty good idea of what TPC is designed to accomplish—an environment for project work in which everyone, from the level of individual contributor working on the smallest activity right up to the CEO, is striving toward the goal of maximized value. All project decisions, such as product features, schedule, budget, and staffing levels, should be analyzed on the basis of quantified data that should be generated right up front in the business case. These will show, at the project level, what makes the project more profitable, and, at the portfolio level, what makes the organization more profitable (which, of course, may sometimes mean canceling even a profitable project).

At the project and activity level, management should be concerned primarily with using available resources to optimum efficiency, whereas at the organizational, or multiproject level, where decisions regarding staffing levels and cash flow issues tend to be made, profit should be the primary concern. This does *not* mean that senior management can ignore efficient resource use, nor that the project and activity managers should ignore profit. What it *does* mean is that, whereas senior management should manage toward profit by using the formula

Projected profit = \$EMV – \$cost,

project and activity managers should manage for efficient resource use as reflected by the DIPP:

Project DIPP = \$EMV divided by \$Cost ETC

and the net value-addeds of activities:

\$NVA = \$value-added – (\$drag cost + \$resource costs)

In the following chapters, we will see how these data should be input, analyzed, and put to use.

chapter four

Planning the scope

Once the Total Project Control (TPC) business case has been developed and reviewed by the customer, product champion, senior management, and/or project management review committee, it reaches the first "gate." A gate is a decision point in the project process. At this point, it will be either killed, tabled, returned for further information and analysis, or approved for detailed scope development. Approval should bring with it funding for the necessary resources to complete *the entire planning process*, including detailed scheduling and costing. Figure 4.1 shows the first three phases through the completion of the detailed plan.

The first gate approves funding through the end of Phase 3 in order to avoid delaying the project. However, the second gate, after Phase 2 when the detailed work scope is developed, can "close" if the work scope document is not approved, thus canceling further funding for Phase 3.

With the business case approved, it is time to assemble a planning team. This is essential for a project of any significant size or complexity. The wide range in the types of work involved in a large project underscores the need for distributed expertise. The precise composition of such a team depends on the nature of the project. However, there are certain functions that are often overlooked. Marketing is probably the most important of these, but others include documentation, training, finance, and maintenance or support, the latter of which will have to live with the product long after its delivery and the disbanding of the rest of the project team.

The job of the planning team is to develop a document that specifies and describes all the components and subcomponents to be developed by the project. This document is variously called the scope document, the product definition, the technical specifications, or the statement of work (SOW). Often these documents include information other than about scope, such as target delivery date, target budget, or a list of available or required resources. This is all nice-to-know information. The delivery date, budget, and resource information will be worked out later in the planning process.

The central purpose of the scope document is simply to define, completely and comprehensively, the intended deliverables.

As we have mentioned before, the scope is the most important part of the project. It is the reason we are doing the project. It is what the *customer* wants from the project. In this regard, by emphasizing the scope to

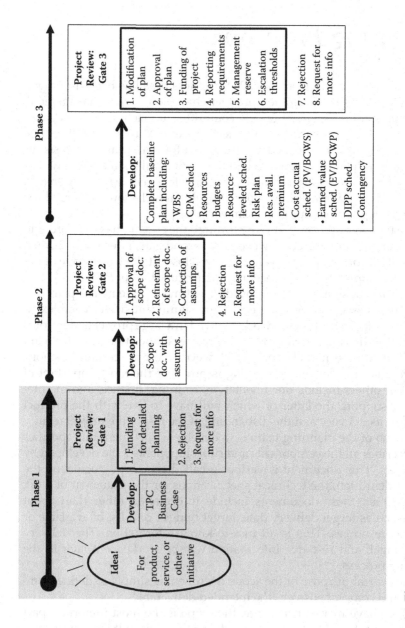

Figure 4.1 The project review process for gating and funding.

such an extent, TPC is in complete harmony with modern business theory by putting the customer first. If the customer does or does not want the product, or does not want a particular component of the product, that should be the driving factor of the product scope definition.

Now, just who the customer is depends on the project. An internal project may have a senior manager as its customer. A fixed-price contract for a specific client will have a specific customer. A product for the retail market may have millions of customers. In all cases, though, a specific representative of the customer must be identified, and must have authority over work scope definition and the work scope document. In the case of the retail market product, this individual should be the product manager or other marketing representative.

I have seen this become a problem. A southern California toy company with which I have done a good deal of work has always suffered from a tense relationship between the marketing managers and the project people (technically, it does not have project managers in the true sense of the word). The people responsible for doing the work felt that the marketing managers were constantly tinkering with the product definition and scope, with no accountability for the schedule delays and costs thereby incurred. The marketing managers, of course, felt that they were responsible for the final product and wanted to make sure that it conformed to their customers' desires. The problem is a classic one in product development projects, when project and product are seen as separate rather than parts of a whole system.

This, of course, is one of the main problems that TPC is designed to solve. Expected monetary value *is* a whole system index, incorporating both the market and the project work. If the toy company had been tracking expected monetary value (EMV) and expected project profit (EPP) via the DIPP (Devaux's Index of Project Performance), each change in the project plan would have been captured and measured in scope value, schedule, and cost, and the DIPP would have forced changes to be justified by data showing an increased EPP. The marketing manager for whom additional project work and costs are invisible will naturally insist on any scope change likely to increase sales and EMV. However, if he can see, and must account for, the cost of the trade-off on schedule and cost (and, further, the opportunity cost through decreased resources and time for other products), he will be less cavalier in calling for enhancements.

The scope document

Defining the scope is by far most important part of the project planning process. If an organization planned all of its projects' scope detail and did no other planning, it would still be better off than the situation that currently exists at most corporations. Companies, and project managers, omit even this fundamental planning step. They invest millions upon millions of dollars in projects where they have only the sketchiest idea of what they are going to do, and, later, of what they have done. Why do they omit it? *Because it is, by far, the hardest and most labor-intensive part of the planning process.*

But, oh, how important it is. Suddenly, the work gets done not by whim but by decision. The scope document can be distributed to the project team, who can ask questions, make improvements, and anticipate what they will need to do, and, of course, schedules, resources, and budgets can be computed, improved, and communicated.

The only question is: How *do* you plan the scope?

Each type of project is different, and each project is different. Therefore, it is difficult to set hard-and-fast rules for assembling scope documents. The best idea I have found is to start with the benefits you want to achieve, incorporate them into the business plan, then move as rapidly as possible to a concrete image of the thing that will provide these benefits. By "concrete," I mean a sketch, drawing, blueprint, model, or prototype, or any combination thereof. Sometimes the deliverable may be particularly intangible, perhaps a service instead of a product. A sketch or flowchart of how the service would be processed and delivered could be of great help. For a software project, the first question should always be: What are the issues or problems this software is being designed to resolve? The second question should be: What screens, data fields, algorithms, reports, etc. will help it to resolve those problems? And third: What will the visible manifestation of the product *look* like? The better defined this becomes, the more efficiently the software can be coded. The interim steps, of incremental bits of functionality that are iteratively expanded through an agile development process, should be guided by the sense of the goals of the final product. That way there is a much greater likelihood that each step (or *sprint*) will advance the effort than if the development team simply takes a tunnel-vision view of the next iteration.

The scope document should be a detailed, comprehensive, and written list of the final deliverable or deliverables to be generated by the project (although, of course, this can all be modified as the project is implemented). It is not until each deliverable has been properly defined that it becomes possible to collect the other necessary data:

- How will each component be designed, developed, assembled, tested, integrated, and whatever else needs to be done?
- Who will be responsible for each phase, component, and activity?
- How long each activity will take?
- What resources will be needed to perform each activity?

- How much will each component, subcomponent, and activity cost, and how much will each component and activity contribute to the expected monetary value of the entire product and project?

The scope document is developed during the initial stages of the project. However, it must be updated throughout the project, because scope is added or subtracted based on the decisions of the customer, product manager, senior management, or project manager. This is no different from agile or other product development processes, but both the vision of the final product and the flexible formats of traditional project management (and TPC) keep the project under much better control than is the case in their absence.

The impact of any scope changes during the project should be analyzed and receive the necessary approval, and then the original scope document amended to reflect the current status. The amended document then should be distributed to all project team members and other interested parties, and adjustments for present and future work, schedule, and cost incorporated into the plan.

The fused memo

However, if there is no such formal process, it is *still* crucial to ensure that the correct scope is being undertaken. The review process must be undertaken by an ad hoc group of reviewers, consisting of the customer, product manager, product champion, senior management, and, perhaps, functional managers and vendor representatives. In such a context, the project manager should distribute the scope document as an attachment to a "fused" memo. A fused memo is one that "goes off" after a certain time, and time is now crucial. For the project plan to be finalized, we must have a defined work scope. It is impossible to "plan to a moving target." This does not mean that once the work scope document has been finalized there will be no further scope changes. However, it does mean that, in order to plan schedule, resources, and budget, we *have* to be able to take a snapshot at some point so as to finalize the plan. Changes to the scope after that point must be understood as changes to the entire project plan, requiring modification of schedule and budget.

The fused memo should, first, call attention to this scope document and assumptions appendix as representing the planned product, and then set a specific date (perhaps two to three weeks hence) after which work scope will be "frozen" and scope changes managed as formal changes to the project baseline plan.

Of course, in the real world, we know what is likely to happen to our scope document, fused memo or not. It is likely to sit in the "in" box of 90 percent of the recipients for weeks. Therefore, the wise project manager will take the time to call the unresponsive managers a couple of days before the fuse expires. Getting the scope correct is *too important* to allow it to be torpedoed by a manager who takes his project responsibilities too lightly.

Whether subject to a formal or ad hoc scope review process, the project manager and planning team must press ahead in developing the plan as quickly as possible. In the formal product development "gating" process, funding for the full planning process through Gate No. 3 was approved at Gate No. 1 precisely so as to avoid unnecessary delays. Similarly, in the ad hoc process, planning should continue even while the fused memo is smoking in the "in" boxes.

"But," comes the rhetorical question, "how can you continue planning when you don't even know what the work scope is going to be? Won't that mean we'll wind up undoing lots of work we've already done?"

There is the beauty of the planning process. For what is the next planning document our team will assemble? Why, the work breakdown structure, of course. And what is the primary function of the work breakdown structure in a project plan? *To help in managing both product and project scope changes.* That's right. The work breakdown structure (WBS) is the project document that puts both scopes into a format that is specifically designed to make it relatively easy for the project manager to adjust the plan to changes in scope.

So, here we find ourselves, as a planning team, with a scope document that is (1) based in substantial part on assumptions, and (2) undergoing review and refinement for the next week or two. So, what do we do? We develop precisely the tool we need to manage all the changes that we know we are going to have to make, and that tool is the work breakdown structure. This project management stuff is great, ain't it?

Appendix A: Assumptions

How do you make plans, or estimates, when you are not sure what work is required? This is a dilemma in which project workers often find themselves. Under such circumstances, the project worker *must* tie any estimate of time or cost to the scope of work he anticipates will be required. If he does not know what that scope of work will be, he *must make assumptions*. Assumptions will allow detailed planning and estimating. However, each assumption and its associated estimate or estimates must be itemized, so that the fact that there is uncertainty about their inclusion will be underlined, and so that when assumptions turn out to be wrong, the associated time and cost can be adjusted or eliminated from the plan.

The assumptions appendix should be a standard part of any project scope document. Indeed, it should be the very first item following the main body of the document. Also, attention should be called to it by whatever means possible, e.g., having the heading printed in red. It is the *most important* part of the scope document in that, by definition, it is where there is the greatest likelihood of confusion and the greatest risk of either desired work being omitted or unwanted scope (and its wasted time and cost.) being added.

As mentioned earlier, defining the scope document is the hardest part of project planning, and the most important. It also should be the most time-consuming. However, we do not want it to be any more time-consuming than it has to be. Many project planners abandon the planning process precisely because it seems to take so long to nail down all the requirements. Doing so is both unnecessary and unproductive. Developing the scope document should be accomplished as swiftly as possible, so that the rest of the planning process can begin. An assumptions appendix, documenting all uncertainties regarding scope inclusions and exclusions, can be a tremendous time-saver, providing a prototype so that others with an interest in the project can refine the work scope.

Once the scope document and attached assumptions appendix have been developed, they must be distributed to those who can check, amend, and delete work scope items that are not wanted and include whatever additional scope is needed. If there is a formal product development process, such as in Figure 4.2, we have now reached Gate No. 2, requiring formal review and approval of the planned work scope.

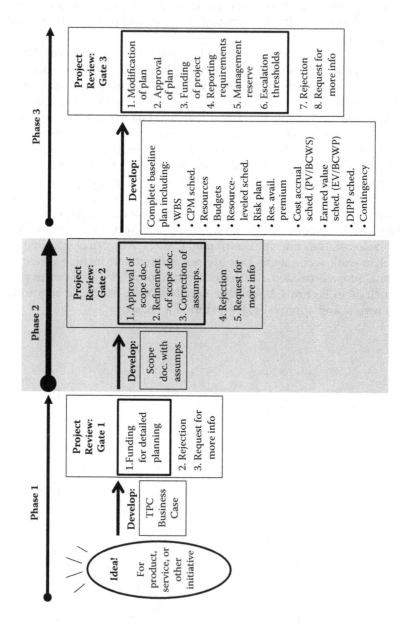

Figure 4.2 Phase 2 of the project review process for gating and funding.

chapter five

Developing the work breakdown structure

If I could wish but one thing for every project, it would be a comprehensive and detailed work breakdown structure (WBS). In organizations with immature project management processes, the lack of a good WBS probably results in more inefficiency, schedule slippage, and cost overruns on projects than any other single cause. When I am brought in as a consultant to perform in the role of "project doctor," invariably there has been no WBS developed. No one knows what work has been done, nor what work remains to be done, and the first thing I have to I do is assemble the planning team to teach them how to create a WBS.

In general, during the years since the first edition of *Total Project Control* was published, the efforts of the Project Management Institute (PMI) have contributed significantly to the understanding and usage of the work breakdown structure. However, the WBS remains the framework or skeleton upon which the entire project rests, and those organizations still not using it are falling farther and farther behind. Remember, scope is the most important part of the project, and the WBS is that scope organized into a detailed hierarchical format. It is the WBS, as we shall see, that ties all three sides of the project triangle—scope, schedule, and cost—together.

Unfortunately, even when a WBS is assembled on a given project, it is often not done very well, due to the planners' lack of knowledge of both the underlying principles and the benefits of a WBS. Often one sees a project manager sitting in front of a computer screen and trying to create a list of activities, formatted somewhat haphazardly into a hierarchy. This list, rarely more than 50 to 100 items long, then becomes "THE WBS" for the project, and scheduling data are superimposed upon it. The fact that dozens of activities, representing weeks of work and tens of thousands of dollars, were overlooked by the project manager becomes evident later on, sometimes just soon enough to convince everyone that "this project management stuff doesn't work."

The WBS is far too important to trust to the efforts of one individual. Because a project usually consists of a wide variety of types of work, and, therefore, requires distributed expertise, that expertise has to be assembled during the most important phase of the planning process, right up front when you are planning the scope. Capturing all the planned project work in the WBS is crucial. Activities that are omitted will not be planned, not be scheduled, not be resourced, and not be budgeted. One could almost wish that they won't be done, either. However, chances are that they have to be. So their absence will be discovered at some point when completed work has to be undone in order to fix the omission, and the result will be schedule slippage and dramatic extra cost.

The project manager *can* develop the upper levels of the WBS hierarchy, but, ultimately, he will get down to a level where greater subject matter expertise than his own is necessary in order to plan the details of *how* the work must be done. At that level, the "branches" of the WBS must be assigned to such subject matter experts, who become the "activity managers" or "project leaders," responsible, and accountable, for planning and performance of those areas of work. In this way, the WBS represents not just an organizational chart of the *work,* but also an organizational chart of those responsible for the work.

The OBS and the WBS

Figure 5.1 shows what, in project management parlance, is called an organizational breakdown structure (OBS). In common corporate terminology, of course, it would be called an *org chart*. From the project viewpoint, it lays out the hierarchical organization of the resources that are available for project work.

The work breakdown structure is similar, in that it is also a hierarchical format. However, the WBS organizes not the *resources,* but the *scope,* both product and project, that has to be done on a project. Figure 5.2 shows the upper levels of a product-based WBS for MegaProdux, Inc.'s MegaMan development project.

The top, or summary, levels of the WBS are the individual components or subcomponents of the product scope. If the project manager can develop these levels and bring them to the initial WBS planning meeting, it can provide a good starting point. If not, the planning team will have to develop these levels.

It has been my experience that there are two good ways for a planning team to develop a WBS. One way is to start at the top, at the project name level, and work their way down by asking what the components of this level are comprised of until the WBS is fleshed out in more and more detail. The other approach, which I have often found successful, is to simply have a brainstorming session:

- First: "What are the components, artifacts, or other tangible items we are going to have to generate in the course of executing this project?"

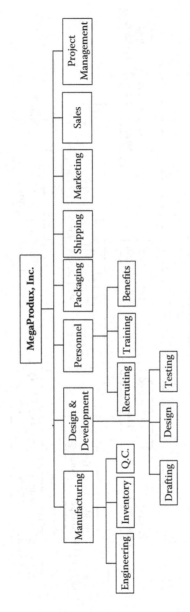

Figure 5.1 Organizational breakdown structure "org chart" for MegaProdux, Inc.

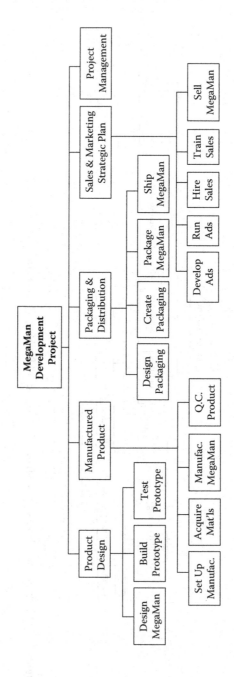

Figure 5.2 Work breakdown structure for the MegaMan development project.

- And, then: "What is the *work* that we have to do in order to create each of those items we just said we have to generate, and to ensure they meet requirements?"

Each item is written on a sticky note and, after a reasonable number (50? 100? 500?) have been generated, the team starts grouping them together—work activities under their respective components in the branches of the WBS. During this process, other work activities will be identified as it becomes clear that "stuff" is missing.

The items at the lowest level of the product scope are called *work packages*. These are products, components, or artifacts, things usually identified by nouns. They are not themselves work, but they "contain" the work necessary to create them. Many people, including the fifth edition of the *Project Management Book of Knowledge (PMBOK) Guide* (PMI, 2013), assert that the lowest level of the WBS is the work package level. This depends on definition; the work package is certainly the lowest level of the project scope. However, each work package is ultimately decomposed into the work activities necessary to create it. Surely that work is an integral part of the *work* breakdown structure? In that sense, the WBS decomposition is not suddenly truncated at the work package level, but continues down to each item of work in the activity list and schedule. Indeed, how do we identify all the work activities if not as the tasks that must be performed to create each work package?

At the lowest level are the *detail activities*. These are the work items, the project scope. These are where *all* of the project work gets done. These are what get scheduled, and these are where resources get assigned. *No* work gets performed at any level above the lowest level.

There are usually intermediate levels between the top of the project and the work packages, and between the work packages and the most detailed activities. They are called *summary activities*, and they serve three purposes:

1. They are a *means* of reaching the lower level, as we break each upper-level component or activity into the more detailed items which comprise it.
2. They are summaries of the lower-level activities, or "buckets" into which the information from the lower levels is poured. This means that summary reports can be prepared and printed based on the information at the summary levels.
3. They can be set up as cost accounts where the budgetary and cost accrual information from the activities below is captured and tracked. (Scheduling information needs to be controlled at a lower level than cost information. Thus resources, whose availability can greatly impact scheduling, need to be injected at the detail activity level, even though the *cost* of those resources can be managed at a summary level.)

Functional versus product WBS

Traditionally, the upper levels of the WBS have often reflected the org chart of the company. This means that the summary levels are designed to be the functional areas: engineering, manufacturing, marketing, etc. Increasingly, however, as companies have recognized the cross-functional needs of projects and have become more "projectized," the traditional WBS has adapted to these trends by also becoming more projectized, with the summary activities becoming components of the ultimate deliverable rather than functional departments. Figure 5.3 and Figure 5.4 show potential functional and product breakdown structures for the same automobile development project.

The product WBS not only incorporates the advantage of presenting "the big picture" in terms of how the project will be done, but also allows a better view of schedule and cost information by component. A customer, a project manager, and a portfolio manager all should have greater interest in seeing project data formatted through a product WBS than through a functional WBS. Additionally, in a moment we will be discussing the value breakdown structure (VBS), and then we will see how the product WBS allows for clear comparison of value versus cost. (Kind of important, right?)

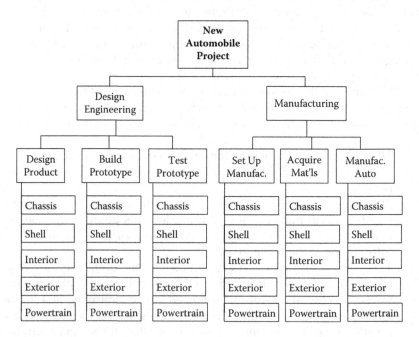

Figure 5.3 Sample functional work breakdown structure for automobile development project.

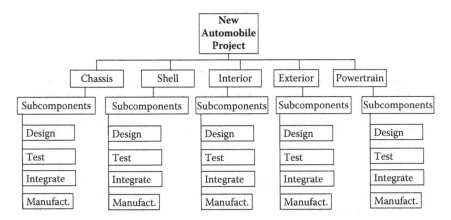

Figure 5.4 Sample product work breakdown structure for automobile development project.

In this regard, the Project Management Institute has been a leader of the movement toward product-based work breakdown structures. For more than a decade, PMI has emphasized its advantages, barely falling short of mandating the product-based approach. However, despite the fact that the product WBS provides the above advantages, this does not mean that a functional WBS is wrong nor that a functional WBS is not *infinitely* to be preferred to no WBS at all.

Indeed, certain areas of a project probably *should* be summarized in a functional format. Work that has been assigned to a vendor, for example, probably should be collected together and reported in a summary of all that vendor's work. Also, it is usually a good idea to keep all project management work on a project under the same WBS summary activity. A WBS that contains elements of both product and functional types is sometimes referred to as a "hybrid" WBS.

Breaking down the WBS

A question that always arises with a WBS is: "How far down do we have to go? How much detail is enough?" And, the answer, of course, is: "It depends." A little later, I shall give some rules of thumb for how much granularity should exist at the lowest level of a WBS, but, at this stage, the planning team is dealing with the initial and upper levels of the WBS. The goal for now is simply to identify all the areas of work so that they can be assigned to subject matter experts who can provide further granularity and estimates. Individual members of the planning team can probably provide substantial further breakdown even at this point. However, on a large or complex project, there will be areas where they will have to

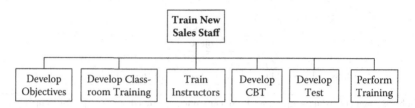

Figure 5.5 Additional breakdown of training activities for the MegaMan development project.

rely on input from others in their departments. For example, the planning team representative from the training department probably knows that training the sales staff will require some classroom instruction, some computer-based training, and a test to make sure that all the students are capable of doing the job. He, therefore, can provide the additional breakdown, as shown in Figure 5.5.

Just *how* all this is to be done will likely require input from the individual instructional designers and computer-based training (CBT) authors who will actually be responsible for the work, and all the other members of the planning team are doubtless faced with similar issues. This initial WBS planning meeting, therefore, should be adjourned as soon as each and every item in the WBS has been assigned as the responsibility of one individual. That individual is now an *activity manager,* responsible for managing that area of the project.

That responsibility starts with the need to get further input from people who may be more intimate with what will actually have to be done. Typically, and preferably, this means those who will actually be doing the work. The activity manager may want to approach each of these workers individually, or may want to organize another WBS planning meeting, only this time to plan that particular branch for which that activity manager is responsible. Whatever approach is used, the activity manager should first review with the workers the planning done at the higher-level meeting, including performing a brief sanity check of the input the activity manager has thus far given to the project manager. Then the activity manager should assist the workers in completing the breakdown of the WBS to the final level of work items or "detail activities" to be performed. The further breakdown on the MegaMan project for the training branch is shown in Figure 5.6.

The activity manager also should begin to prepare for the next step in the planning process by getting estimates of the durations and resource requirements for each detail activity. Additionally, he should collect predecessor/successor information for each detail activity, preferably by assembling a tentative critical path method (CPM) network logic diagram (see Chapter 6) on flip chart paper and sticky notes. All of this information

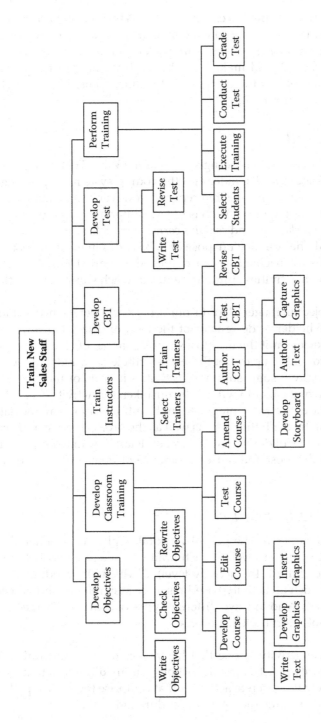

Figure 5.6 Complete breakdown of training activities for the MegaMan development project.

about each activity at the lowest level of the WBS then should be sent to the project manager, either through e-mail or, preferably, entered directly into the project management software package on the intranet. The project manager then should make sure that each member of the planning team receives a copy of the complete WBS, plus, if possible, the preliminary scheduling information.

Coding the WBS

Much is often made of the coding system of the WBS. That is because the software packages are dependent on the coding system for printing the correct information on the desired reports. However, from a management viewpoint, the coding of the WBS is of less importance than the fact that the WBS is *comprehensive* and *contains correct information*.

In general, the primary purpose of WBS codes in most project management software packages is to denote the "parent/child" relationships, i.e., which activities summarize up to which summary "buckets" (Figure 5.7).

Some project managers like to put a period (.) between the numerals, as a place holder or designator of the level of detail. However, most software packages limit the user to a specific number of character spaces, and each period takes up one of these. Nevertheless, periods can be quite helpful, especially if some summary activities have 10 or more children. In this way, Activity 1.4.1.13 will not be mistaken for Activity 1.4.1.1.3.

Having the entire WBS integrated in the intranet allows all the data to be summed up through the code, as well as the project manager or senior management to "drill down" to the level of detail they may need in order to assess and diagnose the cause of slippage or other project anomalies (Figure 5.8).

The detail-level activities

As mentioned earlier, *all* the project work takes place *only* in detail-level activities, or the lowest level on each branch. Therefore this is the level (usually called, simply, the activity level) at which all scheduling takes place and at which all activity-based resource assignments should occur. At this level will be found two different types of activity: discrete activities and level-of-effort (LOE) activities:

- A discrete activity is one that has a specific start and finish. The vast majority of activities in most projects are discrete activities. It is exclusively the discrete activities that comprise the critical path and thereby determine the total project duration.

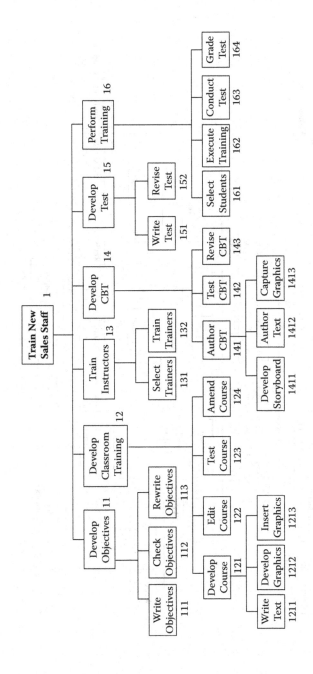

Figure 5.7 Coded breakdown of training activities for the MegaMan development project.

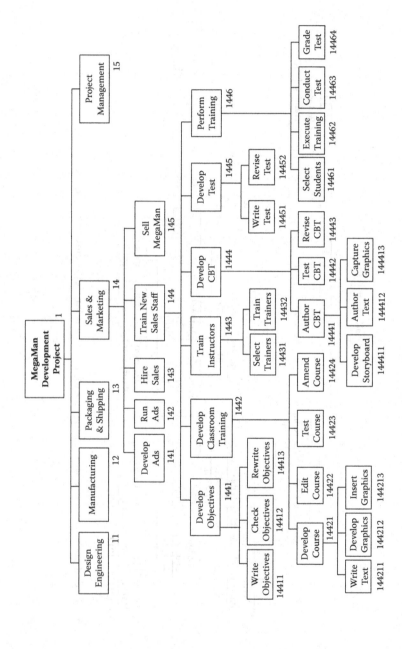

Figure 5.8 Coded WBS with training activities integrated with the rest of the MegaMan development project WBS.

- A level-of-effort activity is one that is ongoing, usually on an on-and-off basis, throughout a portion of the project or throughout the project's entirety. Some examples of an LOE activity include oiling machinery one half-day every two weeks during manufacturing, performing quality control (QC) on a periodic basis, and filing periodic project progress reports. Typically, the start and finish of an LOE activity are triggered by another activity's start and finish, e.g., oiling triggered by manufacturing. The fact that an LOE activity's start and finish are usually tied to another activity's start and finish makes it often referred to as a *hammock*. Therefore an LOE activity should never be on the critical path, although the driving activity that triggers its start and finish often is. The LOE activity's duration expands and contracts in response to variations in its driving activity's duration.

Six guidelines for developing the WBS

The following are six guidelines that are useful in developing a good WBS. I avoid using the word *rules* because it sounds too rigid. Nevertheless, these guidelines are more than mere suggestions. I have seen much confusion generated by WBSs that did not adhere to them.

1. **Use component names that are nouns** (*Outer Surface* or *Durability Test Results*) **and activity names consisting of verb plus object** (e.g., *Paint Outer Surface* or *Test Durability*). First, this will identify the deliverable clearly, and, secondly, it will allow the activity's name to identify both the work taking place and the deliverable to which it is being done.
2. **Each activity should be product-oriented.** In other words, its completion should be marked by some sort of component, a tangible object that unmistakably denotes the completion of the activity. The goal is to make each activity's completion as "binary" as possible. An activity is either completed (product delivered) or still ongoing (product not yet delivered). There should be no argument as to what the "completion criteria" are.
3. **The sum of an activity's "children" must equal the parent.** When an activity is broken down into more detail, each work item planned in the summary activity must be specified in the detail-level activities. Remember, no work takes place above the detail level. Therefore, to omit an item from the detail level that had been intended in the "parent" means not scheduling it and not doing it.
4. **No parent should have an "only child."** This follows from Guideline 3 above. If the "parent" is the sum of its "children," then a single child would mean a redundancy. Get rid of one level or the other. (An exception is if working from a templated WBS where the upper

levels are mandated by procedural requirements. In such cases, "single-child parents" are sometimes required in order to obey the mandated reporting requirements.)

5. **Each activity must be assignable, as an integral unit, to a single department, vendor, or individual.** This is the first answer to the question: "How far down should you break the WBS?" When more than one person is responsible for an activity, *no one* is responsible for it. *This does not mean* that one department cannot use resources from another organization in order to complete a given activity. However, if the work is sufficiently complex that two different departments really *should* be responsible, then break the single activity down to a lower level of granularity and assign each, integrally, to the different areas of responsibility. If each activity is assigned as a separate entity to a different department, then the stage is set for true activity-based costing—each activity can be designated as a project-specific cost account within each department with its own deliverable, budget, and, after scheduling, timeframe. Performance of each department against budget and schedule can be tracked throughout the project, and eventually organization-wide budgeting can be driven by the resources that each department is expending on project activities.

6. **The riskier the work, the greater the detail into which it should be broken down.** This is the second answer to how far down you should decompose the WBS, and it takes us back to the issues involved in the discussion of A–I–M F–I–R–E (see Chapter 2). Detail is one of the key methods of managing risk. The greater the granularity, the earlier it should be possible to identify and isolate a problem area, and to deal with it in an efficient manner.

Two rules of thumb for the level of detail

The level of granularity desirable at the detail level depends on the nature of the project. On refueling outages in most nuclear power plants, activity durations are often measured in quarter-hour units, and schedule progress is reported at the end of every eight-hour shift. Manufacturing processes are sometimes analyzed using activity durations measured in seconds. In either of these cases, common sense tells us that most standard project rules regarding activity durations would not apply. However, for many typical projects, the two general rules listed below can be helpful:

- **Eighty Work Hours.** This guideline sets the upper limit of work effort for any activity at the detail level at not more than two work weeks of effort. By this rule, if an activity is estimated to require more than 80 work hours (or 72 work hours if company procedures assume that each employee spends, on average, four hours a week on nonproject work), then it must be decomposed to a lower level. An organization-wide procedure enforcing this rule can be most helpful in pushing activity managers to provide the level of detail that both project and senior managers sometimes require.

* **One-and-a-Half Times the Progress Reporting Period.** If schedule progress is to be reported every two weeks, no activity should be longer than three weeks; once a month, then no longer than six weeks; and so on. The idea behind this rule is to make sure that no activity ever extends more than one reporting period without a status update, unless it has slipped. In recent years, I have noticed a trend at some organizations toward lowering this threshold to 100 percent of the reporting period.

Again, these are only guidelines. Experience, common sense, and a keen eye in detecting the riskiest areas of the project are the most reliable predictors of what level of detail is needed where.

The WBS as the tool for managing scope change

The WBS is the scope tool. It has always astonished me that this blatantly obvious fact seems to escape experienced project managers and even project management consultants. I have often heard the WBS explained away as: "It's the thing that allows you to get the reports you want out of your project management software." As if it were merely some contrivance developed to respond to the limited functionality of your project management software.

The WBS takes the product and project scope and packages them into nice simple bundles, where the work can be easily visualized and even more easily modified. As soon as the WBS begins to take shape, it begins to help the project manager and planning team crystallize the work that has to be done. The "parent-is-the-sum-of-the-children" guideline allows all members of the planning team to be able to perceive items that have been omitted, often even where they have no personal expertise. For example, in Figure 5.9, it does not take either a genius or a programmer

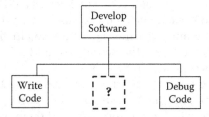

Figure 5.9 WBS fragment for a software coding project.

to figure out that if we are going to *write code* and *debug code*, then at some point we need to *test code*.

In our earlier discussion of the A–I–M F–I–R–E approach, you may remember that the "I" in A–I–M stood for "Isolate"; once the project manager becomes "Aware" of a variance from the plan, she must try to "Isolate" those areas of the project that are not affected from those where the variance has occurred. The WBS allows her to do precisely that, because of the "firewalls" that exist between each work item in the WBS format. In the WBS for the MegaMan project, as shown in Figure 5.2, a schedule or cost variance in the training portion of the project can be isolated from the prototyping, the QC, the marketing, the packaging, etc. (Of course, if there is spillover, which should be captured in the scheduling or costing processes, as we will see in upcoming chapters.)

Perhaps most important is the WBS's role in managing scope change. Adding or subtracting scope is as easy as inserting or deleting boxes in a WBS. For example, if we decide at some point to eliminate using new sales staff, or prototyping, or QC, we simply remove those boxes from the WBS. If scheduling and resourcing have already been done, the schedule, resource, and cost impact of those boxes also would disappear. In other words, remove scope items from the WBS and the effect of those items also will be removed from the other two sides of the project triangle.

This is one of the reasons that greater granularity in the plan means less work to manage change during the project. It is easier to remove three or four small WBS items than to trim and adjust a large one.

It is also the reason that any project paradigm or template for repeatable projects (or fragments thereof) should be stored in the format of a WBS. Rarely is the work scope of one project identical with another. However, they may be very similar. It is relatively easy to take the WBS from a similar previous project, trim the items that are not needed, add new ones that are, and have a WBS for the spanking new project *which contains the actual data for schedule and cost collected from the previous project.* This can represent a huge benefit to companies that do repeatable projects. I remember in 1990 I worked with a toy company that had a "flagship" product, a doll (let's call her Chrissie) that was supposed to be a fashion model and came with such accessories as handbags, mirrors, lighted ramps, etc. The company developed a 1,080-activity WBS for this project, covering design and development, marketing, packaging, and distribution, everything but manufacturing. The WBS was further refined as the project was implemented, and the actual cost and schedule information for each activity was collected in the WBS.

By the following year, the market had changed. Desert Shield gave way to Desert Storm, and the TV was full of Stealth fighters and Patriot missiles and soldiers in desert fatigues, many of them women. Little girls, the toy company's marketing department decided, no longer coveted fashion model dolls, but rather F-16 pilot dolls. And so Chrissie changed. Her hair was cut, her skin tanned (Saudi Arabian sun), and her beautiful gowns exchanged for flight suits or fatigues. The handbag, mirrors, and lighted ramps were replaced by M-16s, grenades, and helicopters.

But what about the WBS? Well, data related to the replaced accessories were now useless. New activities had to be developed for the tools of Chrissie's new trade, and activities related to hair length or skin tone might have to be modified. Perhaps as many as 750 of the 1,080 activities from the original WBS were unchanged, and each of them now had *actuals* in terms of data, not the estimates that had to be relied upon the first year.

But what about *unique* projects? Surely no WBS template can help with such a project? Well, is there really such a project? I worked once with the aerospace research arm of a major university. They were interested in assembling an activity-based costing system. I pointed out to them the advantages to be gained by capturing and storing actual cost data in a WBS template, and reusing them when planning later projects. Initially, they dismissed the idea because "each satellite we put up is different, so all our projects are unique." But what about the telemetry software? The hardware? The communications equipment and network? What about the testing activities? What about activities to transport the satellite to the launch site? What about storage at the launch site? By the time we were finished, we had agreed that more than 60 percent of the activities they performed on every launch not only were *not* unique, they had been performed again and again on launches in the past. However, no one had ever thought to capture and store the actuals in an easily accessible and reusable format like the WBS.

The value breakdown structure

The value breakdown structure (VBS) is a TPC concept that brings the scope/cost/schedule triangle of value analysis down to the project or activity level. Introduced in the first edition of this book, the idea suddenly seemed to catch on, largely in Europe, about a decade after it was first published. The reason may be that it acts as the logical extension of a business analysis technique called MoSCoW that is often used with agile software development techniques. The M, S, C, and W in MoSCoW stand for Must, Should, Could, and Won't, and they represent ways of prioritizing customer requirements. The VBS extends this into the traditional WBS format and monetizes the comparative value-added, even of the optional (i.e., not Must) products and work.

It starts with the quantification, taken from the TPC business case, of the expected monetary value of the project. The purpose of the VBS is to push this quantification down to the level of the components and subcomponents of the project, where daily decisions are usually made without ever taking into account the relative value of the different types of work being done, or the impact of such work on the project duration.

In every project, there is work that is mandatory and work that is optional. Work may be mandatory for two different reasons:

1. The project may simply make no sense without a certain component or activity. For example, if our project is to build an airplane, we must have wings, landing gear, and a propulsion system. Now, *precisely* what the nature is of each of these components, and how they are designed, built, and tested, is optional. Wings may be longer or shorter, straight or swept back; landing gear may be wheels in a tricycle design, skis, or pontoons; propulsion might be by three jet engines, one nose propeller, or a bicycle contraption. But *some* component *must* be created, or our "plane" will be worthless, and have an EMV of approximately zero.
2. There may be a standard, set by a governmental authority or senior management, that no project will be performed without the inclusion of certain work scope. Activities that are required by procedural standards are, by definition, mandatory.

The first step, then, is to determine which components or work items are mandatory, and to assign such activities a value equal to that of the entire project, because the project cannot be performed without them. In the VBS in Figure 5.10, each such work item has been given a value of $10 million, or 100 percent of the current project EMV.

Next we move to the optional activities. Of each of these, we ask the question: "What would be the value of this project if we performed all other activities but this one?"

Let's take our prototyping activities as an example. Obviously, it makes sense to group the activities for building and testing the prototype as one item, since you can't test without building, and you build in order to test. What would happen if we go straight from design to manufacturing, without any prototyping? There may be some organizational information available, in the form of a historical database, of what has happened in the past when we have designed and manufactured a product like MegaMan without a prototype. If not, we should spend a few hours talking with a manufacturing engineer and/or marketing manager. What *does* happen under such circumstances?

Well, wherever the information comes from, we learn that when we don't prototype a product like MegaMan, 50 percent of the time it makes no difference. A good manufacturing process is developed anyhow, and schedule, budget, and sales are unaffected. So 50 percent of the time, *prototyping will add zero value to our project.* The project would still have an EMV of $10 million without prototyping.

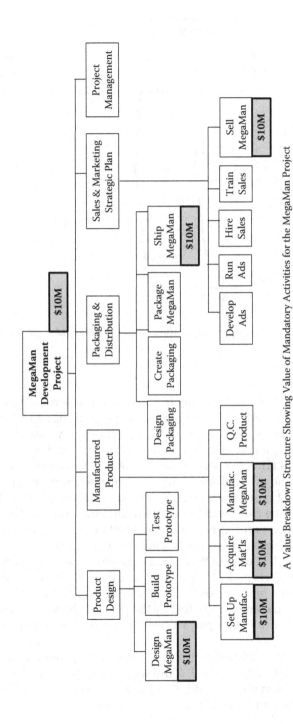

A Value Breakdown Structure Showing Value of Mandatory Activities for the MegaMan Project

Figure 5.10 A value breakdown structure showing expected monetary value of mandatory activities.

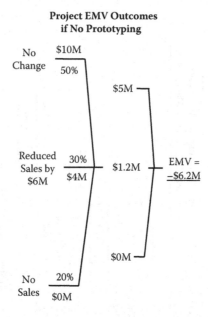

**Project EMV Outcomes
if No Prototyping**

No $10M
Change   50%

                                    $5M
                            Reduced   30%   EMV =
                            Sales by $1.2M   −$6.2M
                              $6M $4M

                                    $0M
            No   20%
            Sales $0M

Figure 5.11 A "tree" diagram computing the value of the MegaMan project without prototyping.

However, 30 percent of the time, there will be problems developing a good process. The result will be delays, scope pruning, and general chaos. From a combination of delays pushing out delivery dates into the actual holiday shopping season, and scope reduction "on the fly" making the product less attractive, our research shows that sales revenues are typically way off the original marketing estimates when prototyping of such products is not performed. Overall, if we don't develop a prototype, 30 percent of the time there will be a 60-percent reduction in EMV. This means that, if we don't prototype, then 30 percent of the time the value of the project will only be 40 percent of $10 million, or $4 million.

The remaining 20 percent of the time, we are *never* able to manufacture a viable product. Rejects make the product worthless. The EMV is zero.

What is the EMV of the project if we don't prototype? It is the sum of the three possible outcomes multiplied by the percent chance of each outcome occurring. Figure 5.11 displays a "tree" diagram showing the value of each outcome.

As shown, a reasonable estimate of the value of the project if we don't prototype will be $5 million plus $1.2 million, or $6.2 million. Therefore, the value that prototyping is adding to the project is the difference between $6.2 million and $10 million, or $3.8 million.

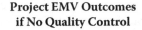

**Project EMV Outcomes
if No Quality Control**

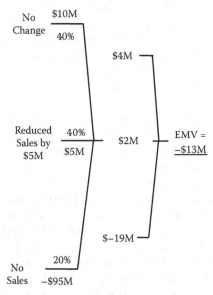

Figure 5.12 A "tree" diagram computing the value of the MegaMan project without quality control.

Let us take one other activity, *Q.C. Product.* It certainly is optional. How much value is it adding? Once again we ask the question: "What would be the value of this project if we performed all other activities but this one?"

Again we refer to whatever historical data exists, and we check with our quality control engineers. The information we gather shows that, 40 percent of the time, rejects will be within acceptable limits even without QC, and the project's EMV will not be impacted. However, a further 40 percent of the time, rejects will rise to a level where customer returns and customer dissatisfaction will lower sales by 50 percent, thus reducing the EMV to $5 million. The remaining 20 percent of the time, not only will the sales be lowered, but also injuries caused by our product will result in a class-action lawsuit that is likely to cost our company $100 million to settle, reducing EMV to negative $95 million.

What is the EMV of the project if we don't perform QC? Again, it is the sum of the three possible outcomes multiplied by the percent chance of each outcome occurring. Figure 5.12 shows the calculation.

As shown, a reasonable estimate of the value of the project *if we do no QC* will be (– $19 million) plus $6 million, or – $13 million. Therefore, the value that prototyping is adding to the project is the difference between

– $13 million and $10 million, or $23 million. Except, of course, that the project is only worth $10 million in the first place. No activity in a project can be worth more than the project itself (or, indeed, more than the parent activity of which it is a part). Therefore, the value of *Q.C. Product* is … $10 million, since we should never perform this project without QC. If we don't perform QC, the value of the project should not be – $13 million, but zero.

As value-added is assigned to lower and lower levels of the VBS, it is good to bear certain things in mind:

- Although the value-added of a single "child" can never be greater than the value-added of its parent, the sum of the values-added of all a parent's children *can* be greater than that of the parent. (Indeed, a single mandatory parent could have several mandatory children.)
- In some cases, the values-added of children can be additive, summing to the total value-added of the parent. For example, if all sales of a product are to be made in one of two ways (e.g., telephone sales and door-to-door sales), then the value-added of these two activities should equal the value-added of selling all the products.
- While it usually makes sense to compute activity value-added as a raw number, it should ultimately be translated into a percentage of the total project EMV. In that way, if the project EMV changes due to market forces or schedule slippage or work scope pruning, the value-added would change accordingly, but remain a constant percentage of the project EMV unless specifically altered.

Figure 5.13 shows what a VBS for all the activities at a certain level of the MegaMan project might look like. Notice that the packaging and advertising activities have been combined, just as the prototyping was. We will refer to this VBS during a later discussion of CPM scheduling.

The value of computing value

How accurate and useful is the estimating and assigning of value? Well, as the cliché goes, to some extent you get out of it what you put into it. Obviously, more analysis tends to generate more accurate numbers, but the fact is that project decisions are made all the time without any attempt whatever to quantify their impact.

Every day, projects are funded, expanded, delayed, and terminated. Resources are hired, reassigned, retrained, and fired. All of this is usually done without any quantitative data being used for purposes of comparison between competing options. Scope is designed, and then trimmed; other activities are added, willy-nilly and on the fly; and when the danger of a slipped deadline looms, resources are added or quality compromised without a scintilla of research to back up the decision.

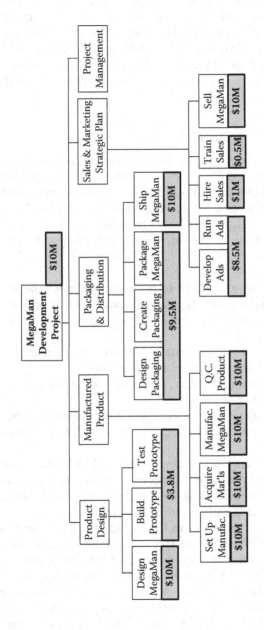

Figure 5.13 Complete VBS for the MegaMan development project.

The analogy it brings to mind is that of a poker player playing "no-peek" poker, where he must bet (read: "invest his resources") without ever seeing any of the cards. Such games are known to minimize the impact of the skill levels of the various players, and to maximize luck. However, "no-peek" poker prohibits *all* players from seeing the cards; no player in his right mind would volunteer to play "no-peek" while his rivals have a free view of the cards, but that is just what happens in the business world every day, and those players who take off the blinders have a *huge* edge.

Estimating and accuracy

As we prepare to move on to our discussion of scheduling, this is a good time to reflect on the issues surrounding estimating and accuracy. Scheduling project work is largely dependent on estimates, and estimates, by definition, are inexact. Efforts to increase the accuracy of project estimates are usually helpful, but also can sometimes backfire, resulting in precisely the opposite of the intended result, namely longer and more expensive projects.

- **The person who is going to be responsible for the work should be the one who generates the estimates.** This is probably the most important contributor to accurate estimates. The reasons for this include:
  - This person will be a subject matter expert, trained in the discipline necessary for the particular work.
  - This person is the only one who will know precisely how he plans to do the work.
  - He will, usually, have a vested interest in meeting his own commitment, and establishing the reliability of his own estimates.

- **The work that is being estimated should be broken down to a sufficient level of granularity to generate accurate estimates.** This will both facilitate estimating, and provide sufficiently detailed milestones to provide early warning of inaccuracies.
- **Estimates that turn out to be wrong should neither be allowed to pass unnoticed nor be treated as felonies.** An estimate represents a commitment from the estimator to make every reasonable attempt to fulfill the prediction. Both individual team members and the entire organization are depending on the accuracy of these estimates. Therefore, an underestimated activity duration that results in schedule slippage should not be ignored. That said, however, we are all human. Sometimes things change, and sometimes we just underestimate what it's going to take to get the job done.

There is no need to act as though this is a hanging offense. "You say you made a mistake in estimating, and your activity is going to take four weeks instead of two? Fine, that's why we have a project plan on the computer. We will input this variance, adjust the schedule, and see what the impact will be on this and other projects, and on our DIPP (Devaux's Index of Project Performance) and corporate bottom line. If the impact is serious, we will make other modifications to come up with the best possible solution. We can anticipate that, with practice, your estimating will become more accurate. Only, *please,* next time have enough detail in your plan to let us know about the slippage before the last minute." The very worst thing that can be done, however, is to punish, or even severely castigate, the individual responsible. Padded estimates are the death of good project management, and the surest way to ensure padding across the entire organization is to crack down heavily on exceeded duration estimates. Haul just one activity manager over the coals for slipping a schedule, and from that day forward every duration estimate on every project in the organization will be padded, totally defeating the purpose of good scheduling—shorter project durations for more profitable projects.

- **Contingency planning at the activity level should be encouraged, with contingency time and cost being in the project plan as separate line items.** Sometimes, depending on very specific circumstances, it is impossible to know precisely how long an activity is going to take. For example, it may be that a new computer-aided design software package has been purchased, which, if it works as promised, will halve the normal amount of time that it has traditionally taken to do a specific activity, from six weeks to three. The activity manager believes there is a 75-percent chance that the activity will take only three weeks. However, there is also a 25-percent chance that the CAD system will not work as advertised, and the activity will wind up taking the usual six weeks. Both duration estimates must be planned for, on a contingency basis, with a 25-percent risk factor for the longer duration, and a trigger mechanism, or fuse, built in one week into the design activity, by which time we will know which estimate is correct. But, and this is key, the 25-percent risk factor of a three-week delay must be comprehended and accepted when estimating the project's EMV and DIPP for initial project funding.
- **Finally, the most accurate predictor of the future is the past. A historical database reflecting the actual data for duration and cost from previous projects can be extremely valuable in the estimating process.** Such databases can be purchased for some applications, e.g., construction. However, databases that have been assembled

from projects performed within the organization itself are likely to have greater accuracy and value. Ultimately, any imported database estimate must be checked against the reality of this specific project context. It's no use pretending you are riding Secretariat if what you've really got is a 30-year-old jackass.

chapter six

Scheduling I: The critical path method (CPM)

Project management is about planning your project in a flexible format so that you can adjust to changes when they occur. Nowhere is that "flexible" format more in evidence than in CPM scheduling.

History of the critical path method

Most project management literature puts the birth of modern project management as 1957 or 1958. These are the years in which CPM and PERT (program evaluation and review technique) were developed in the construction and defense industries, respectively. Today the two terms are used more or less interchangeably. When the boss says to give him a PERT chart of this project, chances are that he is asking for a CPM-derived network logic diagram of the activity schedule, even though PERT actually means something slightly different.

In 1964, IBM project managers developed an "enhancement" of the traditional CPM network logic diagram called the precedence diagram method (PDM). "Enhancement" is used here in quotes because it is arguable if it really represented a step forward. The argument, however, is somewhat mooted by the fact that the method has become commonplace, and is included under the name CPM in just about all project management software packages.

There are two different diagramming techniques used for displaying CPM workflow: activity-on-node (AON) and activity-on-arrow (AOA) diagrams. In AON diagramming, the activity is represented by a box or a node, while the predecessor/successor relationship is represented by an arrow pointing from the predecessor to the successor. Figure 6.1 shows an AON diagram with Activity A as a predecessor of both B and C, and both B and C being predecessors of D.

AON is by far the more intuitive and simpler method of diagramming CPM. With AOA, the arrow represents both the relationship and the activity, as an arrow running between the start of an activity and its finish. The arrow, therefore, is expected to serve two functions, and it cannot always serve both adequately. As a result, it is sometimes necessary to include "dummy" activities in an AOA diagram. These are "activities" that don't

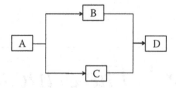

Figure 6.1 Example of an activity-on-node diagram.

really exist and have zero duration, but which must be included in order to properly model the relationships. Figure 6.2 displays the same four-activity project as in Figure 6.1, but, in order to show that both B and C are predecessors of D, we have to include a dummy activity to tie the finish of C as a predecessor to the start of D.

AOA is an obsolete method that is rapidly disappearing as fewer and fewer software packages support it. It is still sometimes used in Europe and, to a lesser extent, Canada, but even there it is rapidly being replaced. Not wanting to support inferior methods, this book displays all logic diagrams exclusively in the AON format.

Using CPM

The fact that CPM is still so neglected in the corporate world more than 50 years after its discovery is a disgrace. Senior managers who complain about projects slipping, yet take no action to enforce good CPM practices, are providing a disservice to their companies.

In order to use CPM for scheduling, you need two items of information about each work activity: duration and precedence. This information should be supplied for each detail-level activity as broken out at the bottom of each branch of the work breakdown structure (WBS) by the activity manager (usually the subject matter expert responsible for the activity's work). With each chunk of the WBS delegated to a department, vendor, or individual, the hierarchy of delegation has been determined. Each

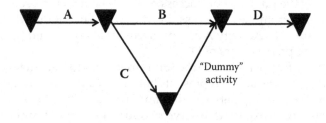

Figure 6.2 Example of an activity-on-arrow diagram, showing a "dummy" activity.

delegate is responsible for supplying the information about the activity that has been assigned to him or her.

Once duration and precedence information have been generated for every detail-level activity, the foundation will have been laid for implementing CPM.

Duration estimates

A duration estimate is for the amount of elapsed time it will take to perform each activity. Depending on the type of work, the time may be measured in units ranging from seconds to months. On many projects with work planned to continue for months or years, estimating in days or weeks is satisfactory. On large maintenance projects with a huge cost of time, such as nuclear plant refueling or airliner maintenance cycles, durations may be estimated in one-hour or even quarter-hour units. In analyzing manufacturing processes, CPM is sometimes implemented in units of seconds in an attempt to identify delays and streamline the process.

Under any circumstances, it is important to remember that duration estimates are always just that, *estimates*. They may be more or less accurate or grounded in historical data, but they always have the potential to be wrong. If they are wrong, the project manager needs to identify and measure the variance and predict its impact as soon as possible. Early variance identification is assisted by using short-horizon estimates. (If an activity is estimated for 10 days duration, it can take 10 days to detect a variance. If, on the other hand, it is broken down into five 2-day activities, variance may show up at the start of Day 3.)

If we know the amount of time it will take to perform each activity in a project, we are in a position to calculate the project's total work time. For example, suppose that we are planning to take an early morning airplane flight to another city. Table 6.1 identifies the activities that must be performed.

The sum of the durations of all the activities is 130 minutes. If our estimates are accurate, this project will take 130 minutes to execute because each activity must be completed before the next one can be started. Such

Table 6.1 Activities to Be Performed before Taking an Airline Flight

Activity	Dur. (min.)
Get ready	45
Drive to airport shuttle	30
Ride shuttle to terminal	15
Check in	15
Go through security	20
Run to gate	5

activities, where each has to end before its successor may start, are said to be *serial activities*.

One way to shorten the project would be to perform some of the activities simultaneously, or in parallel. It is then that the techniques of CPM come into play.

Management reserve, contingency, and padding

There is one more issue to address before leaving our airplane flight project. Assume that each of the duration estimates is a 50-50 estimate: 50 percent of the time the actual duration will be longer than the estimate, 50 percent of the time it will be shorter, and the overs and unders will even out. If we allow exactly 130 minutes to catch our plane, what percent of the time would we expect to miss it? The answer is 50 percent.

For most people, missing one's plane half the time is not acceptable. We, therefore, should add contingency time to the end of the project, a safety buffer that, if any of the six previous activities takes longer than expected, will still permit us to make our flight. The size of the safety buffer, sometimes called *schedule reserve*, can vary depending on such risk factors as time of day or urgency of the trip. A rush-hour drive is likely to require more contingency; the next-to-last flight of the day may have the last flight as a backup or contingency plan, and, therefore, we may feel safe enough to allow less contingency.

There is an important difference between schedule reserve and padding. Schedule reserve is always added either at the end of the project or immediately before a major milestone. It belongs to the project manager and the entire project, and serves as a buffer in case any activity in the project slips. If it is not needed, it should not be used. On the other hand, padding is used for the buffer that a particular estimator may build in to *each activity*. It is the activity manager's slush fund of time or money that is usually added invisibly into the estimate and that almost invariably, by dint of Parkinson's law (*work expands to fill time available*), winds up adding time, increasing cost, and reducing the project's expected monetary value (EMV).

The impact of padding

Perhaps more than anything else, padded estimates are the monkey wrench in the spokes of project work. Estimators pad, that's a fact. The padding, combined with Parkinson's law, turns six-week projects into five-month money pits or worse.

Imagine that you are the activity manager responsible for the *Design Packaging* activity on the MegaMan project. You are asked to develop a duration estimate. You talk to the individuals in the packaging department responsible for each of three subactivities involved in *Design*

Packaging: *Write the Text* that goes on the back of the package, *Typeset the Text*, and *Design the Graphics*. The text writer, Sam, knows from experience that this task will take about eight dedicated hours. He, therefore, might be expected to estimate the task duration at one day.

Sam also knows that, far from dedicated work time, he likely will be required to switch off onto three other projects from time to time. Such multitasking is one of the great wasters of corporate project. Sam knows that writing speedily and well requires immersion in the subject matter and thought processes of the topic. Each time that Sam switches projects, he has to "retool" his brain and this takes time. So, Sam knows it is likely to take him about 12 hours to do this job, say 16, to be on the safe side. However, he also knows that, with the other competing projects, he is not likely to be able to dedicate 16 hours to this project in anything less than two weeks.

Sam also remembers how, late last year, he had a project that he thought would take about 12 dedicated hours and submitted a formal estimate of three weeks. The writing, however, turned out to be more technical and complex then he had anticipated. Then, he had been unexpectedly pulled off to work on a high-priority project. (And truth to tell, the hometown football team had been at a very interesting stage in the schedule.) Despite Sam's best efforts (including unpaid overtime), his one-month estimate had turned out to be a week too short, and Sam had been chewed out by an irate project manager.

That's *not* going to happen to Sam again. If there is even a chance that this project is going to be more complex than it looks, Sam is going to be on the safe side this time. Thus, he gives you an estimate of four weeks.

The estimates you receive from the graphic designer and the type-setter are similarly inflated: five weeks for the graphics and three weeks for the typesetting. Now you have to provide your own estimate for the time it is going to take to generate the overall design and coordinate and check the work of everyone else. Even though some of the work could occur in parallel, you, too, decide to be on the safe side. Your estimate for *Design Packaging* would normally be two months, except you know that the project manager is almost certain to arbitrarily reduce your estimate by a third. (Well, not really "arbitrarily"; all the project managers in the company *know* that everyone pads, so ...) So, in the end, you submit a for-mal estimate of three months.

You were right. The project manager trims your estimate to two months. Two months to do about 40 hours of work.

And, it will take two months, too, because of Parkinson's law. Even if a miracle happens and Parkinson's law is repealed on this project, allow-ing *Design Packaging* to be completed in one month, we would like to be able to start the succeeding activity (*Create Packaging*) a month earlier than in the baseline plan. Unfortunately, this means changing plans, often on a multiproject basis, and that is always disruptive and sometimes impossible. The project duration has been permanently swollen.

Despite the extra time in the schedule, there is still no guarantee that the project will finish on its due date. In fact, it's just the opposite. A corollary to Parkinson's law can be deduced that a percentage of activities will always expand to take *even longer* than the available time. All the padding built into an estimate won't stop the human tendency to do everything at the last moment.

Anyone familiar with corporate project work will instantly recognize the above syndrome. To summarize, projects take much longer than they need to because

- duration estimates are often padded;
- Parkinson's law dictates that activities will almost always take at least their estimated duration; and
- critical path activities taking longer than their estimates always result in schedule slippage, whereas the rare instances of activities taking less time than their estimates seldom result in schedule acceleration.

Further, duration estimates are padded because the prime operative for project work is: *Finish on time. On time* is defined as the expected finish date of the baseline schedule, and therefore everyone goes to great lengths to guarantee that there is more time in the schedule than they need. The paradox is that the very emphasis on meeting schedules causes the project to take much longer than it otherwise would. (Of course, it also can cause disastrous scope reduction or quality compromises.)

Estimating padding

The cost of such estimate padding in terms of reduced EMV is enormous. The pressures that cause padding must be eliminated from any company that is serious about its project work, but accomplishing this is much easier said than done.

First, all estimators must be instructed to provide estimates based on their median realistic expectations. The term realistic is intended to give pause to those wide-eyed optimists who believe that everything always goes smoothly, or that you really can spend each hour of an eight-hour day working. Conversely, it is not intended to reflect what the duration will be if a hurricane, tornado, flood, volcano, and plague of locusts all strike. However, if experienced activity managers believe they will be crucified for not meeting their estimates, either through verbal castigation, negative performance review, or simply being required to work unwanted overtime, they will often submit to the urge to be safe and pad the estimates.

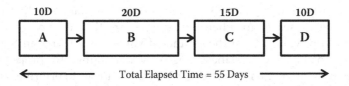

Figure 6.3 Example of an activity-on-node diagram with 25-percent padding built into each estimate.

The habit of padding built up over years of being indoctrinated to meet duration estimates at all costs can only be overcome through retraining. It must be stressed that the baseline plan is not intended to be a precise and accurate prediction of the future. Instead, it is a flexible tool for identifying and adjusting to changes when they occur. If a duration estimate is wrong, so be it; we will adjust, whether underestimation or overestimation. It may not always be possible to draw in the schedule when an activity finishes early, but, if you can, you do. It's a lot easier to do this when everyone understands the potential for change and the benefits of flexibility.

The cumulative padding from all the activities can then be saved and restored at the end of the project schedule as schedule reserve, where it is available if any activity, or path of activities, slips. Figure 6.3 shows a project schedule in which each activity has a 20-percent padding factor built in as a margin of safety. Figure 6.4 shows the same project schedule, but with the 20-percent margins accumulated and stored at the end of the entire schedule.

There are two main advantages to scheduling the project shown in the second diagram:

1. The project team will be working to a schedule from which the padding has been factored out, thus removing the effects of Parkinson's law.
2. If anything in the entire project schedule slips (even activities that are not on the critical path), the management reserve is there as a buffer.

It is important to remember, however, that there is usually a price to be paid for dipping into the reserve. What would be the EMV of the

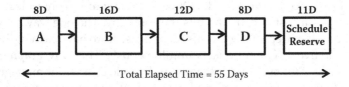

Figure 6.4 Example of an activity-on-node diagram with 25-percent schedule reserve at the end.

project if it were completed 20 percent earlier, without resorting to any management reserve? In other words, what are the delay cost/acceleration premium values from the estimated value/cost of time in the Total Project Control (TPC) Business Case? By managing the project at all times to maximize the DIPP (Devaux's Index of Project Performance) and expected project profit, both project manager and team members will utilize the management reserve only when it is unavoidable, or when the DIPP analysis shows that the delay is less costly than other remedies, such as additional resources or scope trimming.

Working to the DIPP

The best lever for ridding the project of the padding mentality is a paradigm shift. One needs to shift away from the model that stresses the deadline toward one that has the value of the project as its prime operating metric. If team members know that they are working on the critical path and that every day of their activity's duration reduces the project value by, say, $100,000, a new awareness would come into play. The data from the value breakdown structure (VBS) will show the value that each activity is bringing to the project. TPC's drag and drag cost metrics will show how that value is being reduced by the activity's contribution to the project's overall duration. Right from the planning stage, team members will work together to optimize the project schedule, figuring out ways to reduce the duration of their critical path activities. Their goal should be to maximize that item on which they should be evaluated—the contribution of their work to the project EMV. Once implementation begins, the entire team will work to the DIPP-optimized schedule like an orchestra playing a musical score.

The impact of multitasking

A second important factor that contributes to activity durations being needlessly long is multitasking of resources. Anyone who has ever worked in a corporation knows how wasteful this practice is. Yet its impact on project durations and thus on EMV and profits is studiously ignored because department managers want to make sure that their employees always have enough work to do. Otherwise they are likely to lose them the next time a cost-cutting fad hits the company. The one way to ensure that the graphics department keeps all its artists is to have each individual working on five different projects simultaneously. That way, graphics will be a bottleneck for every project and it will be clear to senior management that headcount cannot be reduced. Such situations are particularly common in organizations where critical path scheduling for projects is not standard operating procedure.

Of course, in situations like this, it also becomes almost impossible to justify hiring additional resources, because the precise effect on the project schedules is impossible to measure. Such departments invariably become drastically understaffed. The individuals working in them labor long hours, hopping from project to project in order to keep everyone happy. ("See. Your project's coming along nicely. I'm planning to work some more on it tomorrow afternoon.") Project managers add more and more pressure to hurry up the work, while the department manager, dealing with an impossible situation, reacts by creating a "black hole" department, in which there is so much pressure that not even information can escape. "I don't know when we'll be able to get to this. All my people are working on four different projects. I'll put someone on it as soon as I can, but don't expect it back in less than three weeks."

It is my observation, based on my years of consulting with a wide variety of industries, that every project-driven corporate organization is grossly underresourced in critical functions on project after project. That situation remains unchanged year after year.

Quantifying the impact of multitasking

All this leads to a strong recommendation: Estimators should be instructed to estimate activity durations based on a minimum assignment of at least one dedicated full-time resource, unless decreasing the portion of time assigned will make no difference to an activity's duration. For example, if a quality control person only needs to check the product coming off the assembly line for one hour a day, then there is no need to assume that such a resource should be dedicated to the one activity or project. However, to assign a programmer, for example, to a specific coding job for only four hours a day to a specific project is likely to more than double the length of that task. The estimator must assume that the programmer will be on the job full-time and estimate the duration based on concentrated effort. If it ultimately turns out, when resources are assigned, that the programmer is really only available half of the time, then the activity's duration will have to be adjusted accordingly. It is crucial, then, that such a lengthening of the activity be seen as clearly attributable to the shortage of a specific resource or resources. This is the only way for an organization to start quantifying how much of the time being taken to do projects is due to the way the work has to be done, and how much of the time is due to insufficient or multitasked resources. The first cause of delay is more or less unavoidable, while the latter is often correctable through acquisition of more resources. In a project environment, such additional resources can only be cost-justified by quantifying the cumulative effects of their shortage on projects' delay costs. Without the practice of assuming a dedicated resource for each activity, the inefficiencies of multitasking will remain invisible.

Precedence

A few years ago, while speaking on another subject to a management group at American Power Conversion in Rhode Island, I mentioned the term *CPM* in passing. I was asked to give a definition.

> CPM is a technique for scheduling a project so it takes the least possible amount of time, by doing all the activities at the same time, right up front ... except for the stuff you can't.

My phrasing brought the intended laughs, but it also described exactly the way that CPM is intended to work. CPM allows you to determine the fastest way of doing the project, based on a given set of duration estimates and assuming no resource bottlenecks. The aspect of CPM that tells you precisely what it is that has to come before other work is *precedence* (sometimes called logical dependencies).

Precedence is the logical order in which the activities must occur. For example, in the airport project we must drive to the airport shuttle before we can take the shuttle to the terminal. It is important to note that it is the nature of the work itself that determines the order of the activities in the CPM schedule. We must screw in the hinges before we can hang the door. We must erect all four walls before we can put on the roof. Resource bottlenecks or cost issues will be considered at a later planning stage, not during CPM scheduling.

Precedence determines an activity's place in a project schedule on the basis of two types of activity:

1. A *predecessor activity* is one that must occur immediately before another activity. In the project schedule to catch a plane, shown back in Table 6.1, driving to the shuttle is a predecessor of riding the shuttle bus to the terminal.
2. A *successor activity* must occur immediately after its predecessor. Running to the gate is a successor of going through security.

By arranging the six activities from the airport project into an activity-on-node diagram, the project resembles Figure 6.5.

In the definitions of *predecessor activity* and *successor activity*, note the use of the terms "immediately before" and "immediately after." In the airport project, Activity A is a predecessor of Activity B, but not of C through F. Similarly, Activity F is a successor of Activity E, but not of A through

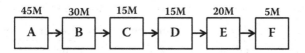

Figure 6.5 An activity-on-node diagram for the airplane trip project.

D. The fact that an item of work must occur before a given activity does not necessarily make it a predecessor in project management terms; it is only the *immediately preceding* and *immediately succeeding* items that are predecessors and successors, respectively. When a project management software package asks the user to enter the predecessors of a certain activity, they only have to enter the immediate predecessors.

Ancestors and descendants

The TPC methodology, however, requires us to define new terms: *ancestor* and *descendant*.

- An *ancestor* is an activity that is on the same path as a given activity, and which must occur before it, but not necessarily *immediately* before it. In other words, an ancestor activity is a predecessor, or a predecessor's predecessor, etc.
- A *descendant* is an activity that is on the same path as a given activity, and which must occur after it, but not necessarily *immediately* after it. In other words, a descendant activity is a successor, or a successor's successor, etc.

By these definitions, all of Activities B, C, D, E, and F in the airport project are descendants of Activity A, and Activities A, B, C, D, and E are ancestors of Activity F. The terms *ancestor* and *descendant* help to denote which activities share the same path and this information will be important in helping to define and quantify TPC's new CPM metric, critical path drag.

CPM logic diagrams with parallel activities

While projects with exclusively serial activities occur all the time in our personal lives (washing clothes and spin drying them, mowing the lawn and bagging the grass cuttings, etc.), they are a rarity in business, where there are usually a variety of resources available to be utilized. In such cases, we can often shorten the total project time by performing two or more activities at once, or in parallel. It is for such projects that the CPM process becomes invaluable, because the precedence is determined by which activities have to wait for other work to be done first. Given the

Table 6.2 Activities, Durations, and Predecessors/Successors

Id	Activity	Duration	Predecessor	Successor
A	Design Product	20D	None	B, C
B	Manuf. Product	40D	A	E
C	Design Packaging	10D	A	C
D	Create Packaging	20D	C	E
E	Package & Ship	15D	B, D	None

logical precedence constraints, CPM helps to determine the shortest possible time the project will take, and what the schedule is for each activity.

In order to explain all the nuts and bolts of the CPM methodology, I will create a project that consists of just five activities. That will be sufficient to show how everything is designed to work without needlessly complicating the picture with other activities. The five activities are shown in Table 6.2.

It is easy to see the sum of the durations of the five activities is 105 days. (For purposes of this example, we shall deal in days. However, in general, we can simply regard the durations as being in generic units of time: hours, five-day weeks, seven-day weeks, months, etc. Whatever is true for days also will be true for any other time unit.)

However, the project does not necessarily have to take 105 days. If, for example, we could start all five activities immediately and perform them simultaneously, the project would take only as long as the longest activity: 40 days. Which activities can be started immediately and which must wait for other work to be accomplished first is denoted by our precedence relationships. We can neither manufacture the product or design the packaging until completing Activity A, *Design Product.* Thus A is a predecessor of B and C, both of which can start as soon as Activity A is finished, but not a minute earlier. Similarly, we must *Design Packaging,* Activity C, before we can *Create Packaging,* Activity D, and we must both *Manufacture Product,* Activity B, and *Create Packaging,* Activity D, before we can *Package and Ship,* Activity E.

- Activity A has no predecessors. Such an activity is sometimes called a *source* activity.
- Activity E has no successors. Such an activity is sometimes called a *sink* activity.

With this information, we can chart the workflow of the project. Duration estimates allow us to turn such a flowchart into an AON diagram (Figure 6.6).

This diagram shows two separate paths through the project: ABE and ACDE. Both paths must be completed in order to finish the project.

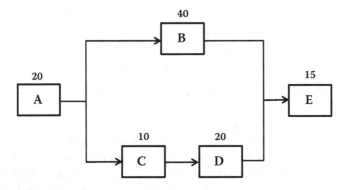

Figure 6.6 An activity-on-node diagram with durations.

Therefore, the project will take as long to complete as whichever is the longest path.

From this comes the definition of the term *critical path*. It is the longest path of activities through the project, and it is critical for three reasons:

1. The longest path determines the length of the project.
2. Any delays on the longest path will make the project longer.
3. If we want to shorten the project, we must do so by shortening the longest path.

Let us add up the durations on each path:

Path ABE = 20D + 40D + 15D = 75D
Path ACDE = 20D + 10D + 20D + 15D = 65D

If this is our schedule, the project will take 75 days and path ABE is the critical path.

The forward and backward passes

All this seems very simple when dealing with just five activities and two paths. However, what if the project were even a medium-sized one, say, 400 activities and 150 paths, how would we figure out the schedule and critical path then?

When I ask this question in my seminars, the answer I invariably get is: "Use a computer." But using a tool when you have no idea how it works is a dangerous road to travel. You will have no idea if the answers you get are the right ones, especially since in this case the term *right* is relative. To get the best results from CPM software, you have to understand what this software is programmed to do.

	DUR.	
ES	ACTIVITY	EF
LS	ID	LF

Figure 6.7 An activity box with durations and schedule data.

Project management software packages contain an algorithm designed to calculate what are called the *forward* and *backward passes*. The output of this algorithm is the traditional CPM schedule with the total project duration. In addition, the algorithm provides crucial scheduling information about each activity.

- The forward pass traces the network logic from first activity to last and calculates the early start (ES) and early finish (EF) of each activity. These are the earliest dates that any activity can start or finish, based on the logic and durations.
- The backward pass traces the precedence logic from the last activity to the first and calculates the late start (LS) and late finish (LF) of each activity. These are the latest dates that any activity can start or finish *without delaying the end of the project* (a key last phrase).

By convention, this information is displayed in each activity box in the manner shown in Figure 6.7.

Formula for the forward pass

The forward pass consists of five steps. The results of Steps 2 through 5 are displayed in Figure 6.8.

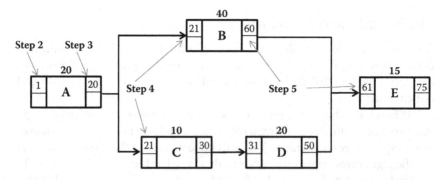

Figure 6.8 Activity-on-node diagram showing Steps 2 to 5 of the forward pass.

Step 1: Assume that each activity will start first thing in the morning and finish at the end of its final day. (Or, for five-day weeks, start at 8 a.m. Monday and finish at 5 p.m. Friday. Or, for hour units, 00:01 and 59:59.)

Step 2: Set the early start for the first activity to Day 1.

Step 3: Compute the early finish for the first activity by adding its duration to its early start and then subtracting 1.

EF = ES + Duration − 1

The reason we subtract is because, by convention, in Step 1 we set the first day of the project as Day 1. Try the formula on an activity that starts on Monday and has a three-day duration. When will it finish? At the end of Wednesday, Day 3, and not Thursday, Day 4, as would seem to be the case if we simply added the duration (3) to the early start (1) and got Day 4.

Step 4: A successor activity can start the morning after all of its predecessors have ended. Therefore, the early start of an activity will be one day after the latest finish of all of its predecessors.

ES = Predecessors' latest EF + 1

In other words, if an activity has more than one predecessor, its start is delayed by the latest finish among those predecessors. Both B and C have only a single predecessor, Activity A, and so can start the morning after A finishes.

Step 5: Calculate the early starts and finishes for the remaining activities by successively applying Steps 3 and 4 of the formula:

EF = ES + Duration − 1
ES = Predecessors' latest EF + 1

Notice how the specific phrasing from Step 4 has relevance when we get to Activity E, which has both B and D as predecessors. Both will have to finish before it can start. Therefore, on the forward pass, it is *the latest finish date of all the predecessors* that pushes out the schedule of the successor activity. E cannot start until the day after both of its predecessors are finished, which is Day 61 due to Activity B's early finish at the end of Day 60.

The forward pass generates the project's duration of 75 days, just as we had computed by adding up the durations on each path. However, the forward pass can be quickly calculated, even manually, for much larger and more complex networks than one would be able to do simply by adding up the durations on each of perhaps hundreds of paths. Even more important, the forward pass provides other vital information—the earliest dates that each activity can start and finish. This tells us the earliest moment that we will need the resources for each activity. There is no point whatever in reserving the resources for Activity D on Day 20; we can't use them until, at the earliest, Day 31.

Now we would like to know what the *latest* is that each activity can start or finish while still allowing us to finish the project in 75 days. This is the output of the backward pass of the CPM algorithm.

Formula for the backward pass

The backward pass consists of four steps:

> **Step 1:** Because we don't want to delay the completion of the project, we start the backward pass by setting the late finish of the final activity equal to its early finish (Figure 6.9).
> **Step 2:** Calculate the late start for the last activity by subtracting its duration from its late finish and then adding 1.
> **LS = LF – Duration + 1**
> **Step 3:** Each activity must finish the day before its successor can start. Therefore, to calculate the late finish for each activity, subtract 1 from the earliest late start of all of its successors.
> **LF = Successors' earliest LS – 1**
> **Step 4:** Go backward through the network, repeating Steps 2 and 3.

Each of Activities D, C, and B has only a single successor and so must finish no later than the evening before the morning that the successor must start. Activity A, however, has both B and C as successors. Activity A, therefore, must finish before either B or C can start. C can start as late as Day 31 without delaying the end of the project, but B *must* start no later than Day 21 if it is to finish by Day 60. This means that A can't finish any later than Day 20, for the project to finish no later than Day 75. On the backward pass, it is the *earliest* start date of an activity's successors that constrains the predecessor activity's late finish date.

With the backward pass completed, we now know the latest that any activity can be scheduled without delaying the projected 75-day duration. This allows for three additional important items of information:

1. We know the latest dates that we will need the resources for each activity.
2. If this were the final schedule, we would know that, if any activity either started or finished after its late date, we would be running late, and remedies would have to be sought in order to get back on schedule. In other words, an early warning system would be in place (as early as Day 21, if Activity A is still ongoing and we

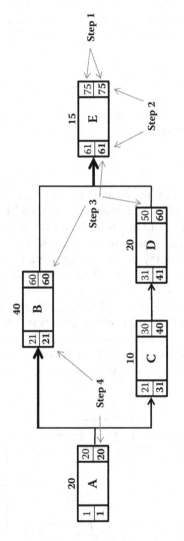

Figure 6.9 Activity-on-node diagram showing each step of the backward pass.

would know that unless something changes the project is going to be late).

3. We have also identified our critical path as
    a. the longest path through the project;
    b. the path where no slippage can occur without delaying the project, because the early and late dates are identical; and
    c. the path where, if we want to shorten the project, we need to change something.

Total float

Total float (TF), also sometimes called total slack (TS), is the quantification of how much an activity can slip without delaying the end of the project. It is calculated using the formula:

TF = LF – EF

Based on this formula, total float for each activity in the product development project would be as shown in Figure 6.10.

Activities A, B, and E all have total float of zero, with identical early and late finish dates. This makes sense, because this is the critical path where any slippage will delay project completion. However, Activities C and D each have total float of 10 days. That means that each can slip up to 10 days without delaying the end of the project. This is useful information to have when assigning resources. Unlike the critical path activities, C and D are somewhat flexible in terms of when they need their resources. They can use them as early as their early starts, but can also wait until their late start dates, which are 10 days later without delaying the end of the project. In this way, the total float metric is used by project management software

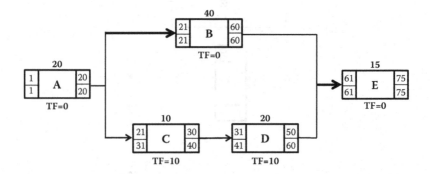

Figure 6.10 Complete activity-on-node diagram of the product development project showing total float for each activity.

to try to eliminate resource bottlenecks during the resource leveling process, which we will discuss in Chapter 9.

It is important to note that total float is not summable along the path. The fact that both C and D have 10 days total float does not mean that each can be delayed for 10 days, up to a total of 20 days. Whatever total float gets used up by Activity C is deducted from the total float available to Activity D. The total float of the path CD is only 10 days.

Free float

Both Activity C and Activity D have 10 days of total float. However, is there any difference between the type of float in Activity C and the type of float in Activity D? In other words, is there any difference in the implications of allowing Activity C to slip versus allowing Activity D to slip?

If Activity D starts on Day 31 (its early start date), but takes an extra 10 days and does not finish until the end of Day 60, does it delay the end of the project? The answer is no. The 10 days are total float and, therefore, by definition, the end of the project is not delayed. Does such a delay in Activity D impact the schedule of anything else in the project? Again the answer is no. Activity E is not scheduled to start until Day 61 anyway. The project is completely unaffected by Activity D's 10-day slippage.

Now imagine that it is Activity C that starts on time, Day 21, but takes an extra 10 days and does not finish until Day 40. Does this slippage delay the end of the project? Again, no. It has 10 days of total float. But does it impact the schedule of anything else in the project?

This time the answer is yes. The effect of Activity C slipping is that Activity D can no longer start and finish on its early dates. The 10-day slippage pushes Activity D to its late start and finish dates. If this were the working schedule, the resources for Activity D might be assigned for Days 31 through 40. On Day 41, those resources might go away, assigned perhaps to a different project. Thus, if Activity D slips beyond its "window" of resource availability, it might wind up slipping even more, delaying the end of the project. So, even though Activity C has total float, any slippage in its dates could delay Activity D and thus indirectly delay the end of the project.

The difference between Activity D and Activity C is that D's total float is of a type called *free float* (FF). Whereas total float is defined as the amount of time an activity can slip without delaying the end of the project, free float (or free slack) is defined as the amount of time an activity can slip without delaying the early start of any other activity (Figure 6.11).

Free float is calculated for each activity by the following formula:

FF (Activity X) = Successors' earliest ES − EF (Activity X) − 1

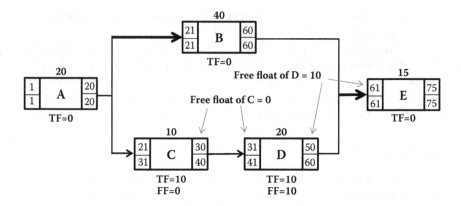

Figure 6.11 Complete activity-on-node diagram of the product development project showing free float for each activity.

Using this formula, we can compare the free float for Activity C with that of Activity D:

FF (Activity C) = 31 (Successors' earliest ES) – 30 (EF of Activity C) – 1 = 0

FF (Activity D) = 61 (Successors' earliest ES) – 50 (EF of Activity D) – 1 = 10

So, Activity D has 10 days of free float, whereas Activity C has none.

Scheduling constraints

Will the critical path activities always have total float of zero? The answer is no. For example, at the start of the project we may get ahead of schedule. In that case, the critical path will have positive total float. Or we may fall behind schedule. In such an event, our critical path will have negative total float (sometimes described as being *supercritical*). In fact, we may have three or four paths with negative total float. All would be supercritical, but which would be our critical path? There are two correct answers:

1. The path whose total float is the most negative would be the longest path to the end of the project and, therefore, would be the critical path.
2. All the paths that have negative total float will delay the project beyond its original planned duration unless we do something about them and, therefore, all may be regarded as critical. All are certainly supercritical.

Both interpretations are valid and useful, but what about before the project starts, at the initial stage of the planning process? When we complete our first forward and backward passes, will the total float on the critical path always be zero?

Not necessarily.

The reason is that sometimes calendar-based considerations take priority over precedence logic. For example, a new union contract, weather risks, or a global meeting may make it desirable that a certain activity occur, or not occur, on certain dates. Most project management software contains the functionality for the project manager to enter date-based schedule constraints that override precedence relationships in CPM scheduling.

There are three general types of calendar-based scheduling constraint:

1. **NET (no earlier than).** This prevents an activity from being scheduled to start (or finish) before a specific calendar date. For example, if the project might require that you sail through the Caribbean in September, you might want to put in a NET constraint to delay the trip until October 15, after the height of the Atlantic hurricane season.
2. **NLT (no later than).** A union contract may expire on June 3, with potential for a strike. In that case, you might want to make sure that certain work gets performed before then.
3. **ON.** This one simply means what it says. The global marketing meeting for this project has been planned for the week of September 15. Flights have been booked, hotel rooms reserved, and so on. This activity must occur that week even if the product isn't at the expected stage of readiness.

What effect does the use of some of these constraints have on the critical path? It could put total float onto the critical path prior to the point of an ON or NET constraint. If our Caribbean sailing activity could start in September, but we are going to delay its start until October 16, every activity on the path prior to that point will have total float added to it. Would it still be the critical path, even if it has total float? I feel that it would. Some software packages, which define the critical path as a path without positive float, would say that the critical path suddenly appears on October 16. However, that path has always been the longest one (and with the constraint, it's even longer), and, therefore, is the critical path throughout the project. Constraints should be considered as part of what can drive a critical path.

Of course, an ON or NLT constraint also can cause a path to have negative total float—without the constraint the path's early dates would have been later.

Using CPM to optimize the schedule

The above calculations of the forward and backward passes are exactly what a software package does when it is commanded to compute a CPM schedule. All too often, the project manager accepts this first version as *the* schedule, and the next step becomes to assign resources (if, indeed, the project manager even bothers with that crucial step, which is too often ignored). This is because she has forgotten that the "M" in CPM stands for method. The reality is that the first network is only the tentative schedule, giving just the data to implement the method fully. Once the initial CPM schedule has been computed and a logic diagram produced, the project manager is in a position to use the critical path *method* to save time and money. If, for instance, we want to shorten the project duration, we now know where to start: the current critical path.

There are two ways of trying to shorten the project:

1. Remove an activity from the critical path. We can do this by finding a way to make the activity no longer dependent on its predecessor. For example, if in the product development project we could find a way of making Activity E no longer a successor of B, we could remove it from the path and thus shorten the project.
2. Shorten the duration of an activity on the critical path. There are three different ways of doing this:
   a. Use more resources (or have them work more hours).
   b. Use different resources (a backhoe instead of 10 laborers with shovels, or a CAD system instead of a draftsperson).
   c. Prune scope. (This is often done, but invisibly and with no computation of the effect it will have on the project's EMV. Pruning scope on a project should always be expected to reduce the EMV. If it doesn't, why was that scope in the plan in the first place?)

But the question is: Where should we add these resources? Where should we cut scope? What will the effect be of such actions on the schedule, cost, and EMV of the project? For the answers, we need TPC metrics.

Critical path drag

If one is using traditional project management software, the data that it will compute for you are those described above. It will give you the project duration, and the early and late starts and finishes of each activity. It will also show you the critical path.

If an activity is not on the critical path, the software will quantify the amount of time by which it is removed from the critical path by calculating the total float of all such activities. Most software packages also will calculate the much less important, but still useful, free float measurement.

What if an activity is on the critical path? What will the software tell you? What kind of quantification will it give you? It will tell you that its total float is zero, which is tantamount to repeating that it is on the critical path. That's all that it will tell you.

Now here's a question. Which is more important: activities that are *on* the critical path or those that are *off*?

The answer, of course, is those that are on the critical path are the critical ones. Yet, when an activity is off the critical path, the software gives you all kinds of useful quantification. However, for the critical activities, the software tells you "*zero*."

Other than quantifying the total duration of the project and identifying the critical path, the *most* important item of scheduling information that the project manager needs is: How much time is each activity adding to the project duration? About this data item, though, almost all commercially available project management software is silent.

Critical path drag (which in the first edition of this book I referred to as DRAG, Devaux's Removed Activity Gauge) is the quantification of the amount of time each activity is adding to the project. It can be thought of as the opposite of total float in that total float is always located off the critical path and is the amount of time an activity can be delayed before its path becomes the longest path (with total float of zero). By contrast, drag is

- only on the critical path; and
- the amount of time by which an activity can be shortened before its drag is reduced to zero (meaning that it is adding no time to the project duration) and another path becomes just as long (and critical).

Alternatively, it is the amount of time that could potentially be saved on the project by removing an activity or by reducing its duration to zero.

Computing drag

Let's look at the network diagram from the product development project (Figure 6.12). Which critical path activities are adding how much time? Alternatively, how much time could be saved by eliminating each activity, or reducing its duration to zero?

Immediately, we can see that Activities C and D are adding absolutely no time to the project; they are off the critical path and each has 10 days of total float (Figure 6.13). Shortening or eliminating C or D would save no time on the project schedule. Conversely, we can save time by shortening any of the critical path activities. How much? Well, if we shorten Activity

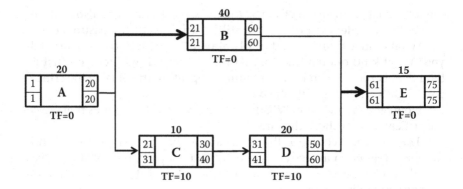

Figure 6.12 Network diagram schedule of the product development project.

A, we will shorten both paths and, thus, the entire project. The same is true of Activity E. However much time we reduce either of these activities by, we will shorten the project by that much. The maximum amount that we can shorten an activity is limited by its duration. You cannot shorten an activity by three weeks if it's only two weeks long. Thus, the drags of Activities A and E are their durations:

Drag of A = 20
Drag of E = 15

Activity B is also on the critical path and has a duration of 40 days; longer than A and E put together. On the surface, it would seem that such an activity would be adding a lot of time to the project, and offer a good opportunity to reduce the total duration. Activity B, however, is

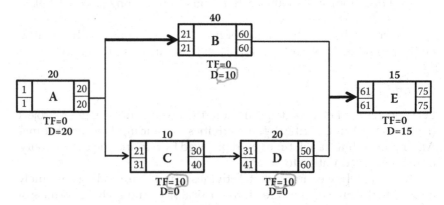

Figure 6.13 Completed activity-on-node diagram schedule with float and drag computed.

actually adding less time to the project than either A or E; it's only adding 10 days. Why? Because as the network diagram clearly shows, once the duration of B is reduced by 10, to 30 days, the project will become 65 days long and the path through C and D will be critical as well (i.e., the same length as the path ABE). If we keep compressing the duration of B all the way down to zero days, we will gain no additional time on the project duration; it will remain 65 days long, with ACDE as the critical path. The additional compression of Activity B will merely cause it to accumulate total float.

What is it that limits the drag, or the amount of time that can be gained, on Activity B? Not its duration of 40 days, but *the total float of the parallel path.* The 10 days of total float on Activities C and D represents the amount of time that can be gained on the critical path activities in parallel with C and D before that critical path changes.

The formula for computing drag on a simple critical path network schedule is as follows:

1. If an activity is off the critical path, its drag = 0.
2. If an activity is on the critical path *and* has nothing else in parallel, its drag = its duration.
3. If an activity is on the critical path and *has other activities in parallel,* its drag = either its duration *or* the total float of the parallel activity with the least total float, *whichever is less.*

To illustrate part 3 of this formula, let us look at an example of a network diagram in which there are not one, but two paths parallel to the critical path, as shown in Figure 6.14.

The inclusion of Activity F, with a duration of 35 days and 5 days of total float, means that if the duration of Activity B is reduced by 10 days, to 30 days, the path AFE will become critical even before the path ACDE can. Activities C and D would still have five days of total float, but Activity F's total float would be zero and now it would have drag of five days. This is why the drag of a critical path activity is limited to the total float of the parallel activity with the least total float, because that is the activity that will next become critical if the parallel critical path activity is compressed.

Now, let us look at Figure 6.15. Activity B has been divided into two activities: B and B'. The sum of these two activities is still 40 days, so the project duration and the total float of activities C and D are unchanged. However, now the duration of Activity B is 38 days and the duration of Activity B' is two days. What will be the drag of the two critical path Activities B and B'?

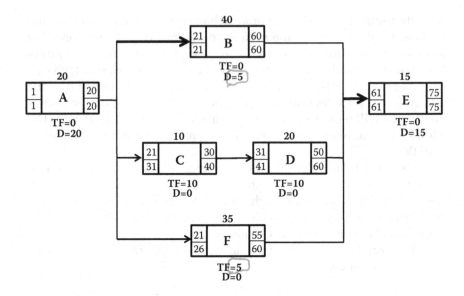

Figure 6.14 Network diagram of a project with two noncritical parallel paths.

The drag of Activity B will remain at 10 days, equal to the total float of Activities C and D, but the drag of Activity B′ cannot be 10 days because its entire duration is only two days. Because its duration is less than the total float of the parallel activity with the least total float, the drag of Activity B′ is limited by its duration of two days. An activity cannot have more drag than its duration.

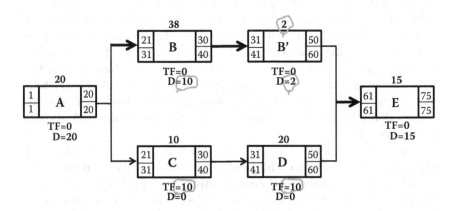

Figure 6.15 Drag with a critical path activity's duration less than the parallel float.

Using drag

The ability to look at network logic diagrams and compute the drag of the critical path activities is vital because of the following:

1. The activities with drag are the ones that are pushing out your project duration, so, in order to shorten your project, you should start by focusing on the activities with the most drag. This is where you can expect to get the most "bang for your buck" by adding resources or pruning scope.
2. Once a project schedule has been adopted and implemented, the drag will be the time that each activity is actually adding to the project duration. As we discussed earlier, time on a project is money, reducing the EMV of the project by delaying completion and delivery. Therefore the drag of an activity has a cost—the amount that the project's EMV is reduced as a result of taking longer to complete. (This, you may recall, was one of the items of information required in the TPC Business Case.) If an activity like Activity B has 10 days of drag, and the TPC Business Case indicates that the project's EMV would be increased by $20,000 for every day earlier that it is finished, then Activity B has a *drag cost* of $200,000.

Most project management software packages do not calculate critical path drag, but this computation is important. A competent project manager or scheduler should learn to compute it through analysis of the network logic diagram.

Let us now try to compute the drag of the critical path activities in a slightly more complicated network, with 10 activities in several paths, as shown in Figure 6.16.

The critical path is ACFIJ, marked by the darker arrows. The other activities all have total float that can be computed by subtracting the early finish date from the late finish date. This yields the total float and drag computations as shown in the diagram.

1. In computing an activity's drag, the first stipulation is that only the critical path activities have drag; so only A, C, F, I, and J have drag.
2. The second stipulation is that, if an activity has nothing else in parallel, then its drag is equal to its duration. Activities A and J have nothing else in parallel and therefore their respective drags are both 10 days.
3. The third stipulation is that, if an activity does have other activities in parallel, its drag is equal to the lowest total float of the parallel activities or its own duration, whichever is least. This means that in order to calculate the drag of activities F, I, and J, we have to determine precisely which activities are parallel with each.

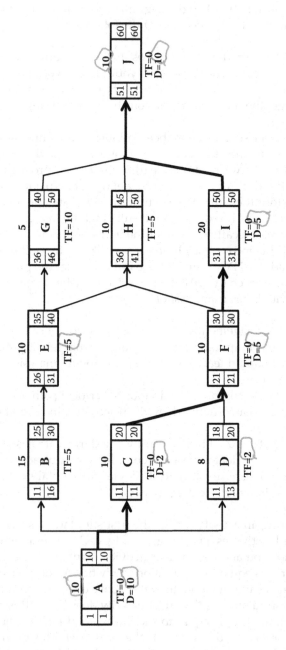

Figure 6.16 Computing drag in a network diagram of a 10-activity project.

Identifying parallel activities is not as simple as it sounds and requires the utilization of the two new terms we defined earlier: *ancestor* and *descendant*. You may recall that an ancestor is a predecessor, or a predecessor's predecessor, etc., any activity that shares the path of arrows and comes earlier. A descendant is a successor, or a successor's successor, etc., any activity that shares the path of arrows and comes later. One activity is parallel to another, conversely, if it is *neither* an ancestor nor a descendant.

Using this definition, let us determine the ancestors and descendants of the Activities F, I, and J. The other activities, by definition, will be in parallel, and the parallel activity with the lowest total float will determine the critical path activity's drag.

- C is a descendant of A and an ancestor of F, H, I and J. By definition, it is parallel with B, E, G, and D. Of those, D has the lowest float of 2. Therefore C, with a duration of 10, has drag of 2.
- F is a descendant of A, C, and D and an ancestor of H, I, and J. By definition, it is parallel with B, E, and G. Of those, B and E have the lowest floats 5. Therefore F, with a duration of 10, has drag of 5.
- I is a descendant of A, C, D, and F and an ancestor of J. By definition, it is parallel with B, E, G, and H. Of those, B, E, and G each has the lowest float of 5. Therefore, I, with a duration of 20, has drag of 5.

Up to 10 days can be gained by shortening or eliminating Activity A or Activity J, whereas only five days can be gained on F or I (despite I's 20-day duration). Also, only two days on C (because of the total float of the parallel Activity D). Therefore, if the project manager wants to shorten the project by adding resources or cutting scope, A and J each offers five times as much potential as C.

Of course, none of this takes into account just what the nature of the work is in any of these activities or whether that work is of a sort that can either be pruned or shortened by adding resources. Some activities are more *resource elastic* than others. (Later, we will introduce a metric for the resource elasticity of an activity: DRED.)

But, remember, the project network that we are analyzing is only 10 activities long; if this were a real project, we might be dealing with 1,000 or 10,000 activities. The critical path itself might be 500 activities long. It therefore would be most helpful to have some simple method of determining where we should start looking when trying to compress the project duration. Activity drags, listed in descending order, provide just such a useful starting point. Armed with that information, the project manager can contact the individual activity managers for those activities with large drags. These activity managers, as subject matter experts, should be able to answer whether or not their activity's duration could be shortened, and what the impact on cost or schedule might be. Through negotiation, compromise, and careful analysis of the impact of the delay cost on the project's EMV, a profit-optimized schedule, as measured by the DIPP, can be generated.

Using drag to recover a schedule

As useful as drag computation is during the planning phase of a project, it can be absolutely *critical* during project execution when things start to slip and the project manager needs to find a way to pull in the dates. This can be particularly tricky in the particular instance where a subdeliverable that is off the original critical path has slipped beyond its due date.

In Figure 6.17, we have "progressed" the network schedule from Figure 6.16. Activity A has been completed on time, but we have just discovered that Activity B will take four days longer than its original estimate of 15 days; it will now take 19 days.

We should immediately notice that our project finish date has not yet been affected. The four days of slippage on Activity B is still within its float of five days.

However, we have included a little wrinkle; our customer has stipulated that the delivery to take place at the end of Activity H *must* occur no later than the end of Day 46. Originally, it was scheduled for the end of Day 45, but now it has slipped out to Day 49. What should we do?

In cases like this (which really are not that uncommon), we have to target the specific activity or milestone that needs to be pulled in. We do this by making Activity H our sink activity and removing all activities that are *not* ancestors of Activity H. Then we recompute the network, the critical path, float and drag amounts just to Activity H, as shown in the numbers in parentheses in Figure 6.18.

With G, I, and J (nonancestors of H) removed, H is our sink activity and, thus, on our critical path. Now we can see that B and E have drag of 9 days and H has drag of 10 days. We now can analyze the work in those three activities to find a way to compress this subset of the schedule by three days and make it so that H can be completed by the end of Day 46.

This is how it should be done, with computation of drag to the target activity, for a network with 10 activities or with 10,000. It just makes the task of figuring out where to look to recover a schedule so much simpler.

Computing an activity's true cost

For all professionals of the project management discipline, it is important to recognize that the vast majority of the members of the finance and accounting professions have little or no understanding of the precise nature of projects and programs nor of the techniques used to manage them. (Indeed, my book, *Managing Projects as Investments: Earned Value*

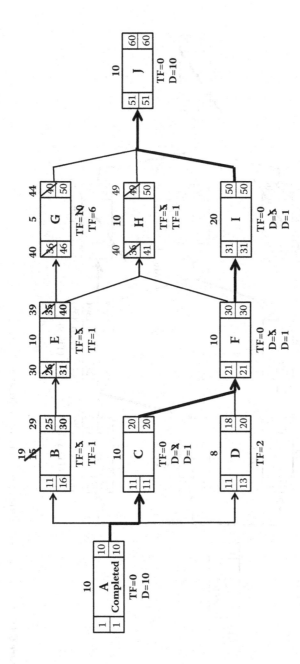

Figure 6.17 Analyzing a schedule with Activity A completed and Activity B slipping.

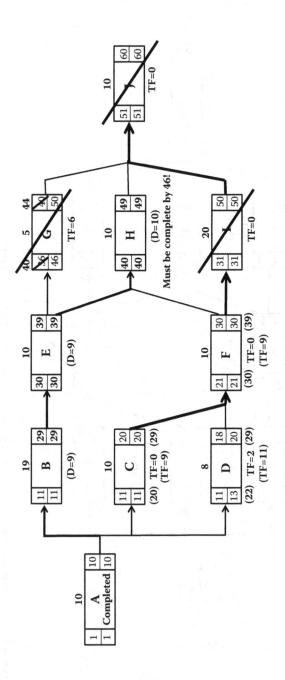

Figure 6.18 Analyzing a schedule to pull in a noncritical activity.

to *Business Value* (CRC Press, 2014), was written with the goal of giving those who see the world through the crucial lens of business value and profit—executives, department heads, and finance, accounting, and business analysis professionals—insight into the essence of projects and project management techniques. The failure of corporate finance departments to monetize and track the value/cost of project time and to relate return on investment (ROI) to critical path management is a major dissipation of both value and resources.)

In over a quarter century of teaching corporate classes in project management, I can count on both hands the number of finance professionals who have been through my seminars. "Why should we attend?" seems to be the sentiment. "What does this critical path thing have to do with finance? A dollar is a dollar, no matter where it's spent."

But by now the reader will understand that this is simply not true. The TPC concepts of drag and drag cost make it very clear that critical path work and resources have completely different implications from those not on the critical path. Both the value and the cost of the project is impacted by, not just the critical path, but the specific activities and resource insufficiencies on that path as measured by their drag and drag cost.

So here is an important insight for project managers: Finance professionals have no concept that the *true cost* of work on the critical path can be vastly greater than for work that has float. From a finance point of view, the cost of work is the overhead-burdened cost of the resources required to perform that work, with no regard to where that work is on the project schedule. As we have seen, however, an activity that has a lot of drag on a time-sensitive project can cost the organization many, many times the dollar value of its resource costs.

The true cost (TC) of an activity is equal to the sum of its resource costs plus its drag cost, and translating that perception for finance and accounting is absolutely key for the project manager who is trying to procure additional resources for her project.

To illustrate, let us assume an activity with a duration of 20 days is on the critical path of a project where the value/cost of time is $5,000/day. Let us further assume that the resources assigned to the activity are costing $2,000/day, or $40,000 over the course of its duration.

Our first step should be to compute the activity's drag. Let us assume that it is 15 days. Its drag cost, therefore, would be 15 × $5,000 or $75,000, and the activity's true cost is $40, 000 + $75,000, or $115,000.

After careful analysis and discussion with the activity leader, it is determined that assigning a second fulltime resource would reduce the duration from 20 days to 12 days. The daily rate of cost would double to $4,000/day, but now for just 12 days or a total of $48,000. The shortened duration also would reduce the drag from 15 days to 7 days and the drag cost to 7 × $5,000 = $35,000. The true cost of the activity, therefore, would become $48,000 + $35,000 = $83,000, or $32,000 less than when only one resource is assigned.

If the project manager decides to implement the new strategy, she may very well push her project $8,000 over budget. However, it's spending $8,000 to increase value by $32,000, and almost any finance professional will see that as more-than-adequate justification, as long as he has been educated about the concept of critical path drag and the value/cost of time on the project has been monetized.

chapter seven

Scheduling II: The precedence diagram method (PDM)

In 1964, an "enhancement" of the traditional critical path method was developed. This enhancement has become so standard today that it has been totally incorporated under the term *critical path method* (CPM), while the term *precedence diagram method* (PDM) has all but disappeared. Today, when a project management software package says that it does CPM scheduling, it really means PDM scheduling.

FS, SS, FF, and SF

One issue with the original CPM was that all predecessor/successor relationships had to be finish-to-start; that is, a successor activity could not start until immediately after the finish of its predecessor. To work around this, project managers would sometimes need to decompose an individual activity into three subactivities: its beginning, middle, and end. By offering new ways of linking activities, PDM seemed to make this process simpler. However, in the adoption of the new scheduling technique, some of the more profound aspects of scheduling may have been lost.

What PDM does is to allow recognition that it is not always the finish of the predecessor and the start of the successor that should be linked. That type of link, called finish-to-start (FS) in PDM, remains the most common and is usually the default relationship in scheduling software.

However, PDM introduces the possibility of three other types of predecessor–successor links:

1. **Start-to-start (SS).** Sometimes it is the start and not the finish of the predecessor that allows the successor to start. The classic example of this is a public works project to lay a new sewer pipe through town. There are two activities: Activity L, *Dig the Trench* beside the road, and Activity M, *Lay the Pipe*. You don't have to finish digging the entire trench all the way through town before you begin to lay the pipe. However, you can't start laying the pipe until you have started digging the trench. The SS relationship would be diagrammed as illustrated in Figure 7.1, with the arrow going from the start of the predecessor to the start of the successor.

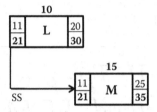

Figure 7.1 Forward and backward passes in a start-to-start relationship.

In a start-to-start relationship:

a. If the earliest that Activity L can start is the morning of Day 11, then that is also the early start date for Activity M. Having determined the two early start dates, we compute the early finish dates in the normal way by adding the duration of each activity and subtracting 1.

b. On the backward pass, we would move backward along the arrows. Let us assume that the backward pass calculations on the rest of the network provide a late finish date for Activity M of Day 35. We would then compute Activity M's late start date as Day 21.

c. Moving backward along the arrow, we go from Activity M's late start back to Activity L's late start and then derive L's late finish. If the latest that M can start is the morning of Day 21, then due to the SS relationship, that moment is also the latest that L can start. And, if the latest that L can start is the beginning of Day 21, then the latest it can finish is 21 + 10 – 1, or Day 30.

2. **Finish-to-Finish (FF).** Sometimes it's not the starts of the predecessor and successor activities that are related at all, it's the finishes, where one activity has to finish before another activity can finish. As an example, suppose that we are going into the citrus fruit business. We intend to truck oranges up from Florida and sell them in Brooklyn. We are also building a warehouse in Brooklyn to store our oranges. We cannot have our trucks of oranges arrive in Brooklyn until we have completed the warehouse. Therefore, we will make Activity N, *Build the Warehouse*, an FF predecessor of Activity O, *Deliver the Oranges*. The FF relationship would be diagrammed as shown in Figure 7.2, with the arrow going from the finish of the predecessor to the finish of the successor.

a. If Activity N starts Day 21, the earliest it can finish is the evening of Day 30. The FF relationship means that Day 30 is also the early finish date for Activity O. Having determined Activity O's early finish date, we compute its early start date as we would on

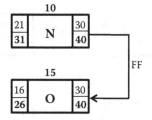

Figure 7.2 Forward and backward passes in a finish-to-finish (FF) relationship.

a backward pass, by subtracting its duration and adding 1 = Day 16.

b. On the backward pass, we would again move backward along the arrows. Let us assume that the backward pass calculations on the rest of the network provide a late start date for Activity O of Day 26. We would then compute O's late finish as Day 40. Moving backward along the arrow, we go from O's late finish back to N's late finish and then compute N's late start.

c. If the latest that O can finish is the evening of Day 40, that moment is also the latest that Activity N can finish. If the latest the oranges can arrive is the end of Day 40, then the warehouse has to be completed by the end of Day 40. And if the latest it can finish is Day 40, then the latest it can start is 40 – 10 + 1 = Day 31.

3. **Start-to-Finish (SF).** This is the most unusual precedence relationship. Instead of the successor's start being dependent on the predecessor's finish, the successor's finish is dependent on the predecessor's start. We organize many aspects of our personal lives according to this type of logical dependency (e.g., we don't quit our old job until we have a new one; many people don't end the old romance until they find someone new). However, there are actually at least two types of important business situations where we should consider modeling them using an SF relationship. These are (1) "just-in-time" inventory control relationships in manufacturing and (2) replacement of any type of legacy system. We could, of course, model each of these using the old finish-to-start relationship with the predecessor activity the one that must occur first chronologically (Figure 7.3). The inventory must be delivered before the goods can be manufactured and the legacy system must be closed out before the new computer system is started.

a. The key to modeling this relationship with an SF link is that we really want the predecessors, *Deliver Inventory* and *Close Out Legacy System*, to be scheduled based on the timing of when their successors will be ready to start. The whole purpose of just-in-time work is to deliver the inventory only when manufacturing

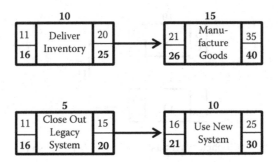

Figure 7.3 Projects modeled with a finish-to-start relationship.

is ready to use it, and we certainly don't want to close out the legacy system until the new system is ready to go online. So, what would happen if, because of delays in the other predecessors of *Manufacture Goods* and *Use New System*, those activities were to wind up slipping by 10 days? As shown in Figure 7.4, because *Deliver Inventory* and *Close Out Legacy System* are predecessors and scheduled to take place before the delay occurs, their schedules would not be affected. The inventory would rust

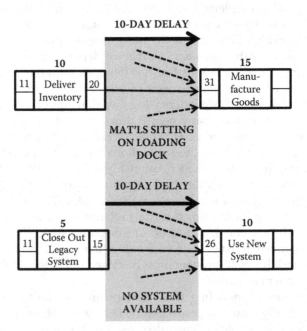

Figure 7.4 Consequences of delays on two projects modeled with finish-to-start (FS) relationships.

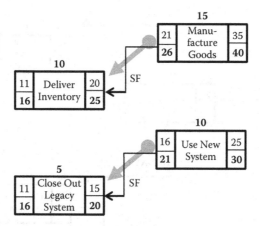

Figure 7.5 The same two projects as in Figure 7.4, now modeled with start-to-finish (SF) relationships.

on the loading dock for an extra 10 days, and for 10 days there would be no system.

Such relationships should be modeled as start-to-finish relationships (Figure 7.5).

b. When modeled as an SF relationship, the driving event (i.e., the start of manufacturing, or the new computer system being ready) can be made the predecessor of the activity that really is dependent in terms of scheduling. If manufacturing is scheduled to start the morning of Day 21, then we need to have the inventory delivered by the end of Day 20. If the new computer system is scheduled to go online the morning of Day 16, then the old system should be scheduled to be deactivated at the end of Day 15.*

c. With the two projects modeled using an SF link, let's see what happens if the same 10-day delays occur, forcing manufacturing out to an early start date of Day 31 or the new computer system to an early start of Day 26. Because the other two activities are now SF successors of where the delay occurs instead of FS predecessors, they too are impacted by the delays. They cannot

---

* Notice that when the two letters, and, thus, the times of day represented, are the same (as in SS and FF relationships), both activities will start or finish at the same time and on the same day (respectively). However, if the two letters are different (as in FS and SF relationships), starts and finishes *always* occur at different times of the day (or week or …), so that the dates will be different—the finish (F) will always be the unit before the start (S). The reader should be aware, however, that some software packages make the finish date of the successor in an SF relationship identical with the start date of the predecessor. This is, in my opinion, just wrong.

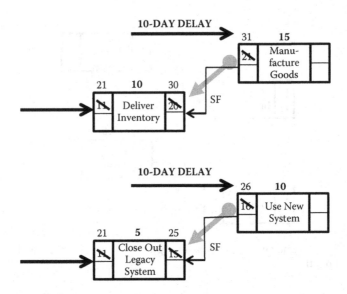

Figure 7.6 Consequences of delays on the same two projects modeled with finish-to-start relationships.

be scheduled to finish until their predecessor is ready to start. Thus they slip out the same 10 days as their predecessors; the just-in-time principle is maintained, and we keep using the old computer system until the new one is ready (Figure 7.6).

In most projects, 70-percent or more of the relationships can be expected to be finish-to-start. About 20- to 25-percent (for those using PDM modeling instead of simply more granular decomposition) are usually SS, and almost all of the rest use FF. SF relationships tend to occur less than 1-percent of the time.

Sometimes two activities need to be tied together by more than one relationship. For example, an SS relationship, such as the one between *Dig the Trench* and *Lay the Pipe*, also requires an FF link between their finishes—we can't finish laying the pipe until we finish digging the trench. Therefore, we also should make the trench digging an FF predecessor of the pipe-laying completion. However, some of the most popular software packages do not support having two relationships between the same two activities. That can usually be worked around by breaking out a milestone (duration = 0) representing the finish (or last nanosecond) of the successor activity. Then the predecessor is not linked twice to the same successor as the finish link goes to the separate milestone.

Figure 7.7 Forward and backward passes in a finish-to-start relationship with lag.

Lag and lead

The precedence diagram method implemented in 1964 also introduced *lag*. Lag is a delay factor injected as part of the relationship between a predecessor and successor.

For example, we may have an activity named *Build Deck* as a finish-to-start predecessor of *Paint Deck*. As everyone who has ever built a wooden deck knows, you shouldn't start painting as soon as you are finished building; you should wait two weeks for the lumber to dry. Therefore, we should inject a delay, or lag, of 14 days into the schedule between the finish of the deck building and the start of the painting. Such a relationship would be called an FS14 relationship and it would be diagrammed and scheduled as shown in Figure 7.7.

Such lags can be built into any type of relationship, as shown in Figure 7.8. For example, we don't want to start laying the pipe at the same instant that we start digging the trench—a length of pipe might roll over a worker's foot. It probably makes sense to spend two days digging the trench before we start the laying pipe. This would then be an SS2 relationship. Similarly, an FF2 link would assure that the successor did not finish until at least two days after the predecessor finished (perhaps providing time to check the renovation work in our Brooklyn citrus fruit warehouse). An SF2 relationship would mean that the successor cannot finish until at least two days after the predecessor starts (allowing us to ensure that we are getting some of the same data out of the new system as we got from the legacy system).

Note that in the last example, the SF2 relationship, the effect of the two-day lag is to cause the two activities to overlap for two full days.

Lag may be input as either positive or negative. Negative lag is often referred to as *lead* and has the effect of subtracting time on the forward pass.

In project management software, the default is a lag of zero. If the network diagram does not indicate a lag value in a relationship, the lag is understood to be zero.

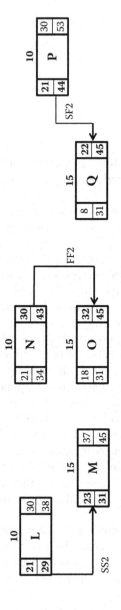

Figure 7.8 Forward and backward passes and precedence relationships with lag.

Two more ways to shorten the project

Earlier, we discussed two techniques for shortening a project:

1. Remove an activity from the critical path by making it the successor of an earlier-scheduled predecessor.
2. Shortening the duration of an activity on the critical path by adding resources or pruning scope.

Precedence and lag now give two additional ways to shorten the project, for a total of four:

3. Change the relationship between a critical path predecessor and successor (e.g., from FS to SS plus lag).
4. Decrease one or more lag values on the critical path.

Adding resources to shorten an activity's duration is sometimes referred to as "crashing the critical path" and it usually requires greater cost for additional resources or reduced project value for pruning scope. Techniques 1, 3, and 4 above, which involve doing more work in parallel, are often referred to as "fast tracking" and tend to increase risk, because new activities are often started on the basis of incomplete information from their predecessors.

Obviously, not all of the above methods can work on any given project, path, or activity. A 30-day test is a 30-day test, no matter how many resources are assigned. You can't debug software code until you have written it.

Where these tools are applicable, however, critical path analysis readily reveals the optimal ways of shortening the project, and, if a Total Project Control (TPC) Business Case as discussed earlier has been developed, the monetary benefit to be derived for such project compression techniques is measurable.

The new product project with PDM

With precedence capability, we can shorten our product development project's planned duration.

- **Change 1:** We don't have to wait until the product is completely designed before we start setting up our lines for the manufacturing process. Twelve days into the design process, we should be able to start setting up for manufacturing. This will make Activity A (*Design Product*) an SS12 predecessor of Activity B (*Manufacture Product*).

- **Change 2:** We don't have to have the entire output of manufacturing before we start packaging. After 35 days of manufacturing, we will have enough product to start getting it ready to ship. This will make Activity B (*Manufacture Product*) an SS35 predecessor of Activity E (*Package and Ship Product*).
- **Change 3:** Due to problems with packaging in the past, senior management has mandated that all packaging designs must be approved by the vice president of marketing. This will cause a one-week delay after the packaging has been designed. This will make Activity C (*Design Packaging*) an FS5 predecessor of Activity D (*Create Packaging*).
- **Change 4:** We don't need to wait until all the packaging has been created before we start the packaging and shipping process. After 15 days of creating the packaging, we would have enough to start the packaging process for any product that has been manufactured. This will make Activity D (*Create Packaging*) an SS15 predecessor of Activity E (*Package and Ship Product*).

The effect of these changes will be to produce the network schedule as shown in Figure 7.9.

The schedule data that are shown in Figure 7.9 are what almost all project management software packages would give us. As we look at the CPM computations, we can see that the project duration is 65 days, and the critical path (with the dark arrows) is ACDE. Activity B, the software tells us, has total float of three days.

However, this raises certain questions about the algorithm that is being used. For instance, total float is defined as the amount of time an activity can slip without delaying the end of the project. What would happen if Activity B slipped out another five days? In fact, if B does not finish

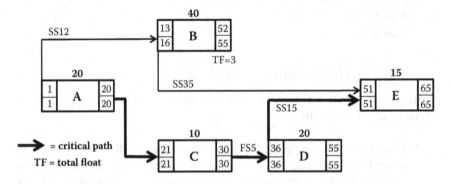

Figure 7.9 Network schedule of new product project with precedence relationships and lags.

until the end of Day 65, it still will not in any way delay the end of the project, at least as it is currently diagrammed.

What about Activity D? It's on the critical path, and reflects this with its identical early finish/late finish dates for zero total float. But what would in fact happen if it too slipped? Couldn't Activity D's finish also slip out by 10 days, to Day 65, without delaying the project's completion?

Clearly, the answer to both these questions, based on the diagrammed dependency information, is yes. Therefore the question we must ask is: Do these relationships adequately model the logic of the situation? Is it, in fact, possible for either Activity B or Activity D to extend until the end of Day 65, and for the project still to finish on that same day? To answer this, we would have to look at the actual nature of the work. Can we manufacture, or create the packaging, up until the last day of packaging and shipping?

The answer, in fact, may be no, but, for the moment, assume that the answer is yes. In that case, the late dates for Activities B and D should be the last dates on which they can finish without delaying the end of the project beyond Day 65, and that is the end of Day 65. If B has a late finish of Day 65, it would still have to start by Day 16 because of its SS35 relationship with Activity E. In order to not delay the end of the project, Activity B must start no later than 35 days before Activity E is scheduled to start, and E must start no later than Day 51. So the work that needs to be completed in the first 35 days of Activity B can only slip by three days. However, Activity B's finish could go out to Day 65, which means that the last five days of B's duration could take an additional 13 days, or 18 days total.

In a similar manner, the start of Activity D is on the critical path and it has to start on Day 36. Its SS15 relationship with E means that the work to be performed in D's first 15 days has to be completed no later than the end of Day 50 so that E can start on the morning of Day 51. However, if the last five days of work in Activity D really *can* slip out all the way to Day 65, then its late finish should actually be Day 65 and it should have 10 days of total float on its finish even though its start is on the critical path.

In a situation like this, almost all currently available project management software would compute the data exactly as diagrammed. That is because it is programmed to calculate the late finish of any activity whose finish is not constrained by simply adding its duration to its late start date.

The very first project management software package I ever worked with was a mainframe number cruncher that allowed the user the flexibility to decide which algorithm to use in doing the backward pass. That software package offered what it called the *slip option*. This meant that if an activity's late finish was not otherwise constrained, it could slip out as far as it would go without delaying the end of the project. The slip option algorithm would have given Day 65 as the late finish for both B and D, as

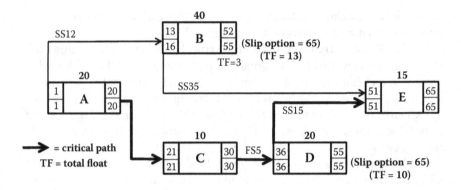

Figure 7.10 Backward pass of the new product project using the slip option.

shown in Figure 7.10. However, most of today's software packages do not include the slip option.

In our specific example, I would suggest that we need to complete both the manufacturing and the creation of the packaging material at least three days before completing the packaging and shipping. We therefore need to add two FF relationships, between B and E and in between D and E, with lag of three days on each. The inclusion of these constraints on the activities' finishes will normally be enough to trigger the algorithms of today's software packages to recalculate the backward pass and to give us the correct finish dates for B and D, as shown in Figure 7.11.

Unfortunately, as mentioned earlier, some of the more popular software packages do not allow more than one relationship between the same two activities. In such cases, we may have to "fool" the software by inputting a final milestone called *Project Completion*, to which B and D are FF3 predecessors and E is an FS predecessor. There is no good reason why we

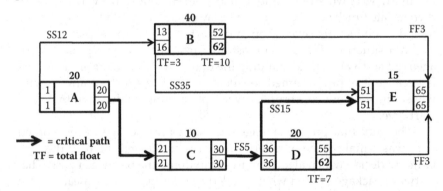

Figure 7.11 Backward pass of the new product project including FF3 relationships to the sink activity.

should have to do it this way, nor why there shouldn't be more than one relationship between the same two activities; this is just another case of software that is produced by designers who don't really understand the full range of ways that their software will need to be used.

Computing drag in a PDM network diagram

Now let's get back to functionality that few of the currently available software packages have—the ability to compute critical path drag. This is an even more important deficiency in PDM modeling because the complexities of SS, FF, and SF relationships, not to mention lag values, make it quite difficult to manually compute drag.

In the PDM network diagram in Figure 7.11, with the FF3 relationships at the end, the critical path is ACDE. However, how much time is each of those activities actually adding to the project? How much time could we save on any one of those activities by adding resources or otherwise compressing the activity's duration?

Think about this for a moment. Here you have a network that's just about as simple as it's possible to get: five activities and four paths. Yet it is a very complex problem to figure out something as basic as how much time each activity is adding to the project duration. Imagine how difficult it is to figure out when you are dealing with even a medium-sized project, say, 500 activities and 400 paths. Most of the software gives you no help with this crucial information.

Because you may not be able to rely on your computer to compute drag, let us see if we can learn how to compute it by looking at the network diagram (which, fortunately, the software will print out for you). In the previous chapter, we learned the following formula for calculating drag:

1. If an activity is off the critical path, its drag = 0. So the drag of Activity B = 0.
2. If an activity is on the critical path AND has nothing else in parallel, its drag = its duration. It's pretty easy to see that Activity C has nothing else in parallel. But what about D? A? E? Partially in parallel? What does that even mean?
3. If an activity is on the critical path AND has other activities in parallel, its drag = *either* its duration *or* the total float of the parallel activity with the *least* total float, *whichever is less*.

So what is parallel with A? Is C's drag equal to B's total float of 10 days? Perhaps surprisingly, the drag totals of each of the four critical path activities are as shown in Figure 7.12.

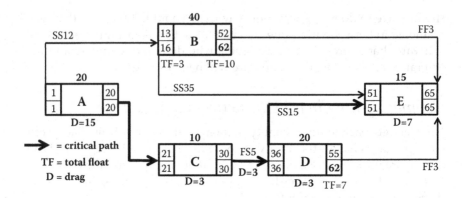

Figure 7.12 PDM network diagram with drag totals for the critical path activities.

Was PDM really an advancement?

Computing these numbers based simply on the above PDM network is tricky for anyone but the most experienced schedulers. The reason is that the PDM relationships make the process of determining parallelism difficult. The truth is that PDM designations are not indispensable for modeling such relationships. Long before the PDM relationships were conceived in 1964, project managers understood only too well that relationships other than simple finish-to-starts existed. But, they modeled such relationships in the finish-to-start format by decomposing activities to a finer level of detail and by incorporating more milestones to represent the start and finish of activities.

It's not as though from 1957, when CPM was first developed, until the 1964 development of PDM that construction project managers dug trenches for 20 miles along the highway before going back and laying the first length of pipe. Managers understood that the relationship between these activities would allow them to proceed in parallel. So they just modeled the activities differently—by breaking the trench digging into two different activities: *Dig First 20 meters of Trench* and *Dig Rest of Trench*. In that case, *Dig First 20 meters of Trench* becomes an FS predecessor of both *Dig Rest of Trench* and *Lay the Pipe* (Figure 7.13).

In PDM modeling of the New Product Project, Activity A looks like one activity. However, the same relationship with Activity B could be accomplished using only FS relationships by decomposing Activity A into two activities with FS successors at the point in A where the lag allows B to start, as shown in Figure 7.14.

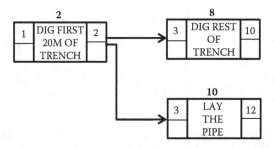

Figure 7.13 A typical start-to-start relationship modeled using exclusively FS logic and decomposition.

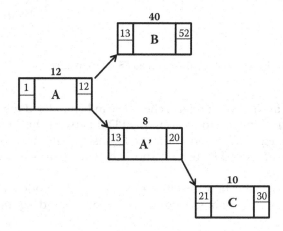

Figure 7.14 Activity A decomposed to model SS + lag relationship with B as an FS relationship.

In this case, Activity A becomes an FS predecessor of both Activity A′ and Activity B, while A′ would be the FS predecessor of Activity C. All SS and SS + lag can be modeled using only FS links in this way.

The effect of an FS + lag relationship, such as between Activity C and Activity D, can be modeled by changing the lag value to what it would have been modeled as prior to 1964 and the invention of PDM's complex dependencies; an activity, whether that activity is *Watch the Lumber Dry* as was the case following the deck-building activity or, in our New Product Project, *Marketing VP Checks Packaging.*

FF relationships can be modeled by creating a milestone (basically, an activity with duration of zero) as an FS successor of both the FF-related activities. Lag with an FF relationship would be modeled as with an FS relationship, as an activity.

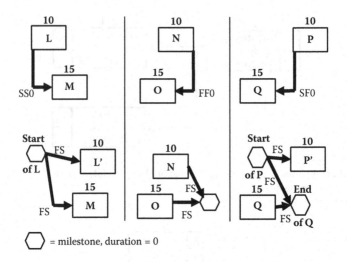

Figure 7.15 PDM relationships modeled as FS relationships.

By these means, any project schedule can be modeled to produce the same schedule that we produced in the PDM network, but using only FS relationships. Figure 7.15 shows how to "translate" all PDM relationships into finish-to-starts and Figure 7.16 shows how to do the same thing when lags are included.

In this way, the new product project can be modeled with only FS relationships, which should make computing drag much easier (Figure 7.17).

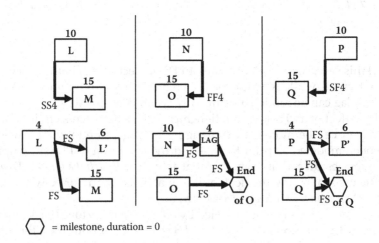

Figure 7.16 PDM relationships with lags modeled as FS relationships.

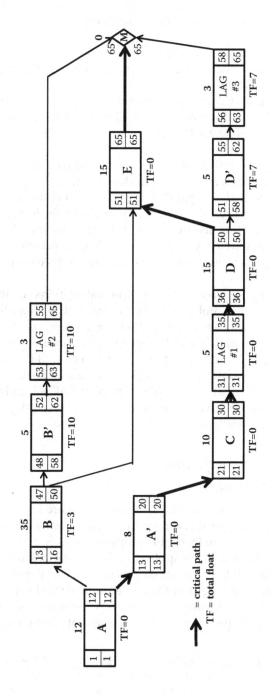

Figure 7.17 The new product project with all complex dependencies turned into FS relationships.

Now we can compute the drag of all the critical path activities using the standard formula that we laid out in Chapter 6:

1. If an activity is off the critical path, its drag = 0.
2. If an activity is on the critical path *and* has nothing else in parallel, its drag = its duration.
3. If an activity is on the critical path and *has other activities in parallel*, its drag = either its duration *or* the total float of the parallel activity with the least total float, *whichever is less*.

The first 12 days of Activity A are on the critical path and have nothing else in parallel. The remaining eight days of A, or A′, are also critical, but have three activities in parallel (i.e., that aren't either ancestors or descendants): B, B′, and LAG #2. Of these, the one with the least total float is B with three days. Thus, the drag of A′ is three days, and the drag of what was the entire Activity A = 12D + 3D = 15 days. Activity A is adding 15 days to the project duration: its first 12 days and 3 days of its final 8 days.

With the decomposed activities, the drag calculations for the other activities are also a little simpler. Activities C and LAG #1 have the same three activities in parallel as A′, and so their drag totals are the same: 3 days. D has been split into D and D′, but D′ is not on the critical path. Therefore, the drag of D is all located in its first 15 days and is equal again to the total float of the parallel activity with the least total float: B = 3 days. Finally, we can now see that D is parallel with B′, D, LAG #2, and LAG #3. Of these, D′ and LAG #3 have the least total float at seven days, so that Activity E's drag is seven days.

A quick method of computing drag in SS relationships

By far the most common relationships in scheduling are FS, but SS relationships are the second most common by a substantial margin. As we have seen, computing drag in FS relationships is relatively easy. However, it would be nice to be able to compute drag in SS relationships without having to decompose all the SS relationships on the critical path.

A few years ago, and long after the first edition of this book was published, I discovered a quick way to do it. I now offer it in this edition for those who are using a software package that does not compute drag:

> The drag of an activity on the critical path that has one or more SS or SS + lag successors (whether the successors are on the critical path or not) can be computed by adding the lag value on each SS relationship

to each successor's total float. The lowest sum of lag and float to an SS successor will be drag of the predecessor unless:

- the predecessor activity's duration is less than that sum; or
- there is a parallel activity with lower total float than that sum.

In Figure 7.19, if Activity X is *not* part of the schedule, then B will have drag of 12 (equal to the lag + TF of Activity F, 5 + 7) and D will have drag equal to its duration of 5. If Activity X with TF of 10 is part of the schedule, B's drag will only be 10. (The drag of D would remain unchanged at 5.)

Notice that this technique works for the New Product Project schedule in Figure 7.12. Activity A is an SS12 predecessor of Activity B, which has TF of 3 on its start (the TF of 10 on B's finish is irrelevant as 3 is less than 12). Thus A's drag would be 12 + 3 = 15, the same number that we got via the decomposition in Figure 7.17.

It is often useful to list activities in descending order of their drag totals, or drag cost totals, or true cost totals. This makes it easy to see the items that offer the greatest opportunity for project improvement right at the top of the list. Additionally, the drag should be related to the drag cost. The drag cost should be calculated based on the amount that, according to the TPC Business Case, the expected monetary value (EMV) of the project would increase if that activity's drag were eliminated so that the project would finish that much sooner. If our project has delay costs of $10,000 per day, the drag costs of the critical path activities in Figure 7.18 are shown in Figure 7.20.

Can we spend $20,000, or even $49,000, to shorten Activity A by five days? It certainly looks like it would increase the investment value of the project by $30,000 or by $1,000.

Generating the CPM schedule for the MegaMan project

In Chapter 5, we produced a work breakdown structure (WBS) and then a value breakdown structure (VBS) for the MegaMan project. That VBS is reproduced in Figure 7.21.

The next step in planning this project would be to assemble an initial CPM schedule. In order to do this, we first meet duration estimates for each detail-level activity. So, let's assume that we have spoken to the activity managers/subject matter experts for each activity and generated the duration estimates.

The next task is to arrange these activities in the order in which they can be performed. This is a task that often can be best accomplished by meeting with the entire planning team. We should write the name of each detail-level activity on a sticky note, and then arrange

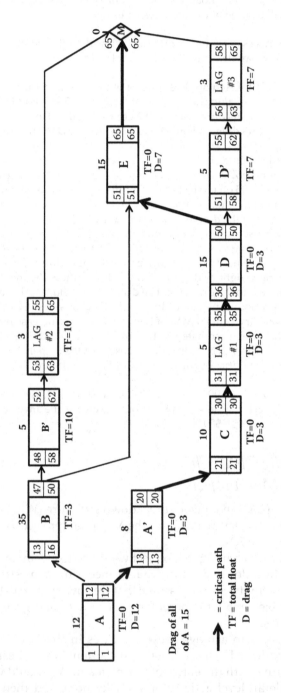

Figure 7.18 The new product project diagram in Figure 7.17 with drag computed.

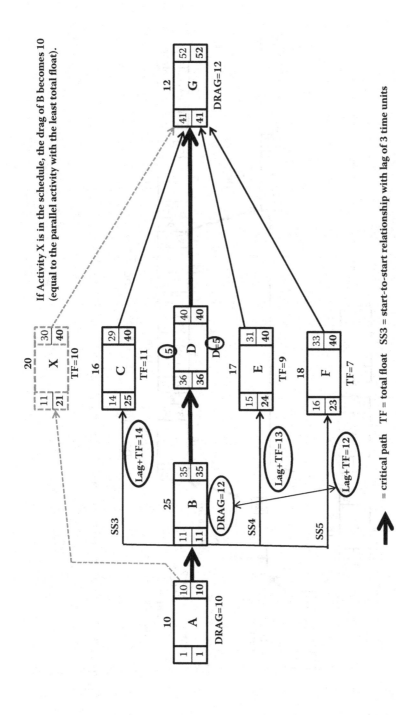

Figure 7.19 A quick method of computing drag with SS + lag relationships.

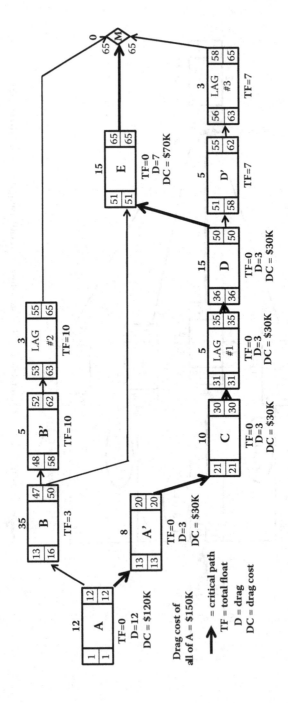

Figure 7.20 The new product project schedule showing drag costs at $10,000 per day.

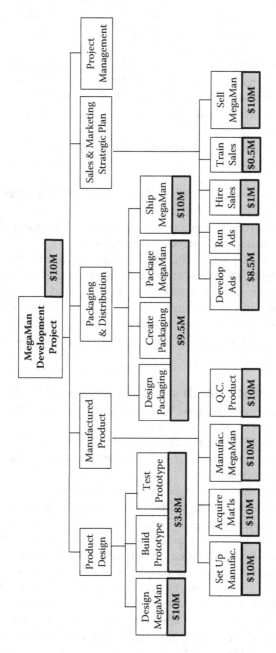

Figure 7.21 The complete VBS for the MegaMan project (from Chapter 5, Figure 5.13).

them all from left to right on flipchart paper with the earlier activities preceding the later ones. Relationship arrows then can be drawn in pencil, and the forward and backward passes computed either by entering the data into a software package or by doing the mental computations on the sticky notes.

Some obvious fast tracking (i.e., SS relationships), such as between *Run Ads* and *Sell MegaMan* or around prototyping, may be helpful, but we don't need to obsess over it too much at this juncture; the schedule is going to change anyway as we optimize it. Figure 7.22 shows the initial schedule for the MegaMan project.

Some of the logic decisions made when inputting this schedule include:

- Complete the design of MegaMan before starting work on the prototype.
- Complete the prototype before the start of both designing the packaging or acquiring the materials to manufacture.
- Include the packaging in the advertising, and do not develop the advertising until after the design of the packaging is finished.
- Hire and train sales staff so as to be prepared to start selling MegaMan as soon as we start running the ads and for at least one week after we finish running the ads.
- Quality control (QC), selling, and packaging will all be completed before shipping begins.

The resulting CPM schedule shows a duration of 40 weeks, but this is only the first pass at a schedule. Is this our best possible schedule? And what does the word "best" mean in this context?

According to the TPC approach, "best" means greatest benefit or most profitable. To determine what our most beneficial schedule is, we must refer to our TPC Business Case.

You may recall that our MegaMan project needed to be completed so that the toy could be in the stores for the start of the holiday shopping season, 30 weeks hence. If completed by that date, the project was estimated to be worth $10 million. For every week later, the value will drop by $2 million. This means that, after 35 weeks, the shopping season will have passed and the project will be nearly worthless. For every week earlier than Week 30 that the project finishes, the value would increase by $400,000. Obviously, the current 40-week schedule is unacceptable. We must find some way to shorten the project. This means examining the critical path, particularly those activities that have the most drag. Figure 7.23 shows the drag totals of the critical path activities.

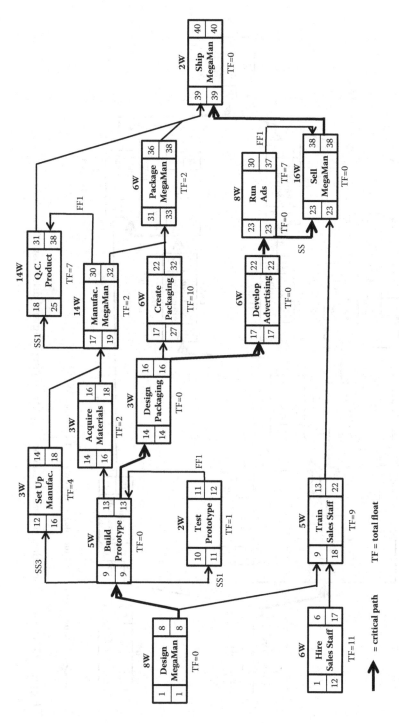

Figure 7.22 The initial CPM schedule for the MegaMan Development Project.

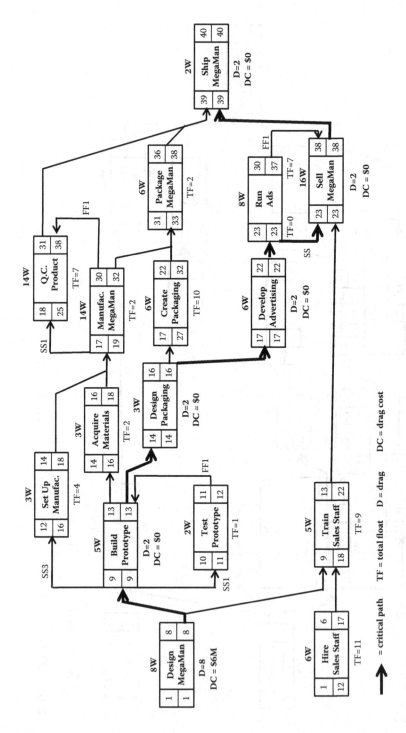

Figure 7.23 The initial schedule for the MegaMan project showing drag and drag cost totals.

In looking at the network logic diagram, we can see the following:

1. *Design MegaMan* and *Ship MegaMan* have nothing else in parallel and, therefore, have drag equal to their durations: eight weeks and two weeks, respectively.
2. *Build Prototype* is parallel to all of *Train Sales Staff* (total float = 9 weeks), *Set Up Manufacturing* (total float = 4 weeks), and *Test Prototype*. Test Prototype has total float of only 1 week. However, since *Build Prototype* is an SS1 predecessor, the first week of *Build Prototype* is not parallel with *Test Prototype*. Using the new method of computing drag for critical path activities with SS successors, the drag of *Build Prototype* would be 2W, or the least of:
   a. 5W, based on its duration;
   b. 7W, based on the lag and total float (3 + 4) of *Set Up Manufac.*; and
   c. 2W, based on the lag and total float (1 + 1) of *Test Prototype*.
3. *Design Packaging* and *Develop Advertising* are both parallel with *Acquire Materials* and *Manufacture MegaMan*, which each have total float of 2 weeks. Therefore, *Design Packaging* and *Develop Advertising* each has drag of 2 weeks.
4. *Run Ads*, despite being on the critical path, has zero drag (0 + 0) due to being completely parallel with *Sell MegaMan*.
5. *Sell MegaMan* is parallel with *Run Ads* (total float of 7W on everything but its first nanosecond) and *Package MegaMan* with 2 weeks of total float, making its drag 2 weeks.

Figure 7.23 also lists the drag costs for each critical path activity. *Design MegaMan* has drag of 8 weeks. If the activity where eliminated from the project or its duration reduced to zero, the project would be 32 weeks long. The TPC Business Case indicates that the 32-week project would have an EMV of $6 million. That is $6 million more than with the 40-week schedule, which is worth zero. Therefore, *Design MegaMan* is costing $6 million because of its drag. We would save $6 million by eliminating its total duration. Compare this to the value of shrinking the *Sell MegaMan* activity from 16 weeks all the way down to zero. There would be absolutely no change in the value of the project because only 2 of those 16 weeks are drag, so that the drag cost of *Sell MegaMan*, and the benefit of having a 38-week schedule instead of one of 40 weeks, is worth zero.

In fact, five other critical path activities each have 2 weeks of drag. If any one of them was eliminated, the project would still have a duration of just 2 weeks less, or 38 weeks, and the project will still have zero value. Therefore, the activities with 2 weeks of drag have no drag cost at the moment. Obviously, just reducing their drag will get us closer to a viable project schedule, but we need to find another place to compress the schedule before any 2-week compression will have significant value.

Using drag to optimize the PDM schedule

Obviously, this information is of great value in determining how to target additional resources. But for the moment, we are trying to shorten the CPM schedule exclusively by fast tracking, or changing precedence relationships in order to do more work in parallel. The drag data is useful for this also.

1. Start with *Design MegaMan* and its 8 weeks of drag. If we start building the prototype before the design is completely finished, we can change its relationship with *Build Prototype* from FS to SS + lag. Of course, we also would have to make sure that the design is completed before we finished building the prototype, so we would need an FF + lag relationship as well. The time saved will be the difference between the current 8 weeks before *Build Prototype* is able to start and the lag value attached to the SS. An SS5 relationship would save 3 weeks.

2. Currently, we are planning to finish building the prototype before we start designing the packaging. This is a conservative approach. By designing the packaging based solely on the design specs for MegaMan, we would slightly increase the risk of a problem occurring in the prototyping process that would force us to change our design. However, *Design Packaging* has 2 weeks of drag that we could save if we succeeded in removing it from the critical path. By making it an SS7 successor of *Design MegaMan*, we should be able to shorten the project by another 2 weeks.

3. The last activity, *Ship MegaMan*, has drag of 2 weeks. However, if we change its relationships with predecessors *Package MegaMan* and *Sell MegaMan* from FS to SS + lag, so that we can start shipping before we finish packaging and selling every last unit, we can save still more time. However, we need at least 1 week after the last unit has been sold and packaged before we can finish shipping. By drawing FF1 relationships between packaging and selling MegaMan and *Ship MegaMan*, we reduce *Ship MegaMan*'s drag to 1 week and, therefore, compress the project schedule by an additional week.

The overall gain from all of these changes is 6 weeks, reducing the project duration to 34 weeks, as shown in Figure 7.24.

The critical path has changed, and so the drag totals have changed. Additionally, with the total project duration down to 34 weeks, the project

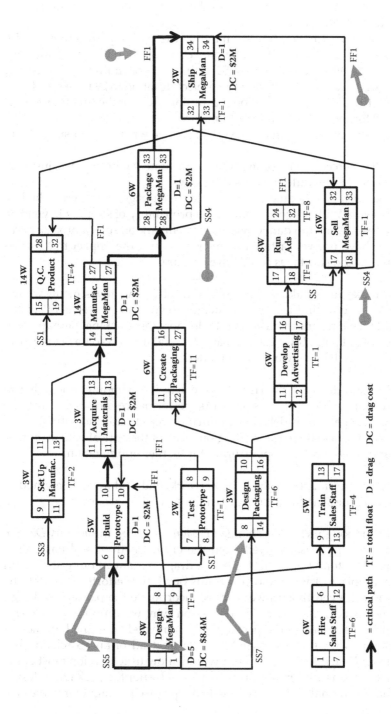

Figure 7.24 Fast-tracked CPM schedule for the MegaMan project, showing new critical path and drag and drag cost totals.

value is now positive, at $2 million, and thus there is now a significant cost for each week of drag.

Eliminating further drag from any of the critical path activities with 1 week of drag would shorten the project duration to 33 weeks and thus increase the EMV by a further $2 million, to $4 million. Eliminating *Design MegaMan* would shorten the project duration to 29 weeks and thus increase the EMV by $8.4 million ($2 million each for the first four weeks, $400,000 for the fifth week) to $10.4 million.

An EMV of $2 million is not very attractive when one considers that:

- a 400-percent increase in profit seems almost within grasp, just four weeks of schedule compression away;
- the current schedule allows for little slippage, with zero schedule reserve. With a tight schedule, the probability of slipping beyond 34 weeks seems too dangerous even to embark on such a project; and
- we have not yet factored in resources, and the impact that bottle-necks and shortages could have on our schedule.

To make undertaking this project an attractive investment, we are going to have to find a way of knocking several more weeks off of the schedule. That means that we need to do more fast tracking, perhaps even off the critical path. If we can increase the total float on some near-critical activities, it will have two beneficial effects:

1. We will decrease the risk of those activities slipping (due either to resource shortages or slow work performance) and becoming the critical path, thus delaying us during project implementation.
2. We may also increase the drag on some of the critical path activities, and thus create the opportunity for further optimization.

So, we continue to analyze the network, and find another method of optimizing the schedule, this time on the second longest path:

3. Even though when we changed *Design Packaging*, the activity *Develop Advertising* was removed from the critical path, it still has only 1 week of total float. This limits the drag on many of our critical path activities to 1 week and constrains our future efforts to shorten the project. Even though we want to incorporate the packaging design in our ads, there is really no reason why we cannot start *Develop Advertising* before the packaging is completely designed; in fact, as soon as we finish *Design MegaMan*. Instead of an FS relationship from *Design Packaging*, an FF1 would allow us to develop most of the advertising earlier, requiring just 1 week after the packaging design has been finalized to incorporate the precise packaging into the ads.

The result of this third enhancement is shown in the network diagram illustrated in Figure 7.25.

This time, although the critical path has remained the same, several of the drag and drag cost totals have changed because of the shortening of the second longest path. Now we have a systematic way to go about trying to shorten the project. We know where the critical path is, and the increased drag values on the critical path give us some room for further shortening. The next step is to involve the activity managers and subject matter experts.

- The first 5 weeks of *Design MegaMan* are costing us $8.4 million in drag cost. What can be done to reduce these 5 weeks and make the project profitable? Can the individuals performing this activity somehow do the work faster to allow the *Build Prototype* activity to start sooner? If we can cut the lag relationship to SS3, we will save 2 weeks and increase our EMV by $4 million.
- Another big opportunity lies in the *Package MegaMan* activity. Its 6-week duration includes 3 weeks of drag at a cost of $6 million. Are we using two shifts for this activity? Three shifts? Do we need to hire more packaging labor? Buy more machines? We've got up to $6 million to spend if we can shorten the activity by 3 weeks.

The two changes above offer the potential to shorten the project by a total of 6 weeks, worth $8.8 million. The first 4 weeks, which would take the project duration to 30 weeks, are worth a total of $8 million. Thereafter, we would be delivering our product to the retail outlets before the prime holiday buying season, and the TPC Business Case tells us that any week gained that reduces our duration to less than 30 weeks is worth *only* $400,000.

This step down in the weekly cost of drag is not at all unusual, but can make things a bit tricky to compute. Take a look at the *Build Prototype* activity. It has drag of 1 week based on the 1-week total float on the finish of the parallel activity, *Design MegaMan*. This is where we must avoid the tunnel vision sometimes inherent in concentrating on each activity, one by one. In fact, if the lag on the FF1 relationship from *Design MegaMan* were removed, the two prototyping activities, if regarded as one activity, would have drag of 2 weeks based on the total float of the parallel activity *Set Up Manufacturing*.

Now let us look back to the VBS that we developed in Chapter 5 and as shown in Figure 7.21 (Figure 7.26).

It is evident that the prototyping activities are *not* mandatory, but they are valuable, to the tune of $3.8 million. However, that is $200,000 less than the cost of 2 weeks of drag if those weeks push the project out to the period between Week 32 and Week 35. If we can do prototyping and still

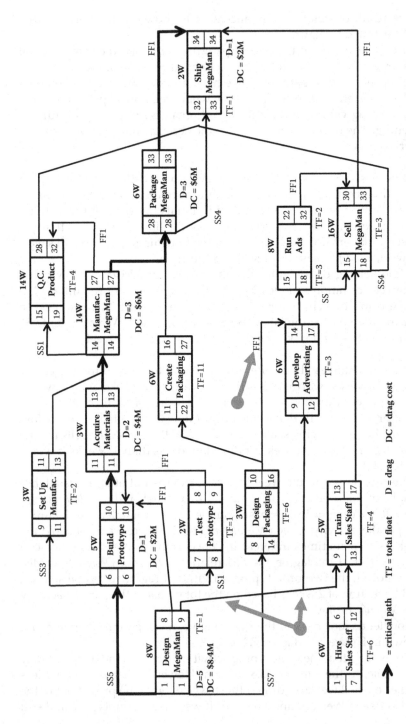

Figure 7.25 Fast-tracked CPM schedule for the MegaMan project, showing further optimization off the critical path and new drag and drag cost totals.

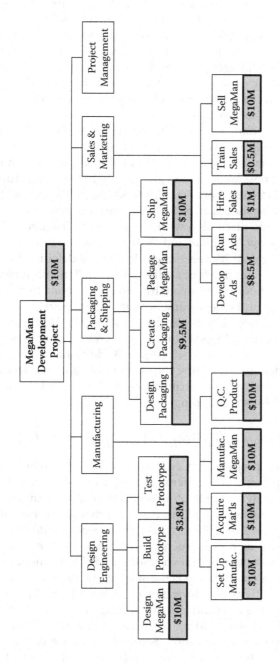

Figure 7.26 Complete VBS of the MegaMan project (from Figure 7.21).

finish the project by Week 30, then those two weeks of drag would only cost $400,000 each, or about $3 million less than prototyping is worth. If the 2 weeks straddle Week 30 (in other words, Weeks 30 and 31), then the drag cost of prototyping would be $1 million plus $400,000, or $1.4 million.

So, whether we should include prototyping depends on what our project duration is projected to be. If we are headed for a completion date of Week 31 or earlier, the prototyping activities will have greater value-added than their drag cost and should be performed (provided the resource costs on the two activities aren't so great as to give prototyping a greater true cost than the value-added). However, if we are headed for a completion date beyond Week 31, the time added to the project duration may be more valuable than the activity of prototyping. We should carefully analyze the possible impact of going straight to manufacturing without the intermediate steps of creating a working prototype and testing it.

These decisions should all be made before we ever start actually doing the project. Perhaps discussion with the activity managers will allow us to shorten the *Design MegaMan* and *Package MegaMan* activities by a total of 6 weeks, giving us a 28-week schedule without deleting the prototyping. But 2 weeks is not much of a safety net on a 7-month project. In our particular case, we don't have much time to decide; *Build Prototype* will be scheduled to start at the beginning of Week 4 and, being a 5-week activity, it will start adding time to the project if *Acquire Materials* is delayed from starting later than Monday morning of Week 7. If we believe there is a substantial risk of a schedule slippage of 3 additional weeks, we should seriously consider either deleting the prototyping and its 2 weeks of drag before we ever start, or even abandoning it after 3 weeks of work if the expected project finish is slipping.

The doubled resource estimated duration (DRED)

In our MegaMan project, with only 16 activities, it is relatively easy to identify the options for shortening drag-heavy activities by adding resources. With only a couple of activities to worry about, we can meet with the activity managers and discuss our options. However, if we were dealing with a project of 1,600 activities, the process would be much more complex. It would be nice in such a situation to have readily available data informing us as to where such resource additions might provide the most impact. Drag computation is the start of this process by quantifying how much each activity is delaying project completion. Once we identify an activity as having a certain amount of drag, however, we still have no way of knowing whether adding resources to that activity will have any meaningful impact on its duration. Some activities are very *resource elastic*—if you double the number of resources, you will cut the duration in half. Other activities may be impervious to resource increases—a 20-day durability test is a 20-day test, no matter how many products are being tested or how many testers are conducting the test.

What we need is some simple way to quantify this quality of resource elasticity. To this end, TPC has developed the Doubled Resource Estimated Duration (DRED). As the name suggests, the DRED of an activity is an estimate of how long it would take if the rate of planned resource usage were to be doubled. For example, digging a trench 100 meters long might be estimated to take four days with a single backhoe. But if we rented a second backhoe (and driver) each day, how long would it take? Two days? Three days? Whatever we determine to be the correct answer would be the DRED of that activity. Another type of activity (e.g., growing a crop of produce) might be completed no faster no matter how many resources we assign; the plants still have to grow and mature.*

Note that the DRED does not necessarily mean that the number of assigned resources has to change. It may just be that the same resources are utilized for more hours. For instance, adding a programmer to a software coding activity could result in confusion, duplicated code, bugs, and schedule slippage. One also could assume that just the one programmer will work longer days and weekends to provide the resourcing level of the DRED. (Obviously, one would have to take into account as well the exhaustion factor on the resources having to work extra hours.)

The DRED does not have to be adopted in its entirety. It is just an easy-to-understand index of the resource elasticity of the activity. If the activity's manager estimates that the duration could be halved by going to the DRED resourcing level, the project manager may interpret that to mean that a 50-percent or 33-percent increase in resources will have a lesser, but proportional effect. In general, there often may be no need to take on more resources than are required to eliminate an activity's drag.

Nor can the DRED be resorted to blindly by the project manager. What the DRED does is allow the project manager, when looking over a schedule of many hundreds of activities, to see those places where additional resources might be added with maximum beneficial effect. The project manager must still check with the activity manager to ensure that the estimate is valid and that a certain number of additional resources will have the desired impact.

The project manager also must establish that the additional resources, in fact, are obtainable. There is no point in dreaming about duplicating a resource if the one we have is unique.

---

* In some cases, an activity can actually be delayed by increasing the resources. There can be an optimal team size for a job, and getting more people involved can simply cause them to get in one another's way. Imagine work being performed in a cockpit or in a submarine. Of course, here again this (negative) DRED information can be vital.

Figure 7.27 shows the DREDs (in parentheses next to the duration estimates) that we have gotten on the activities in the MegaMan project.

A glance by the project manager suggests that meetings with the activity managers for the *Design MegaMan, Manufacture MegaMan,* and *Package MegaMan* activities might be productive. Whether those meetings will result in further shortening of the project duration depends on the specific resource issues. We will cover this in greater detail in Chapter 8.

The reverse critical path anomaly

One brief word of warning about the PDM algorithm as used by almost every project management software package: It can lengthen your schedule through what is called the "reverse critical path" anomaly. (It's very sad when, as happens every week throughout the world on projects ranging from healthcare to national security to emergency response, people die because projects take longer than they could. But, it's particularly sad when the cause of the extra time is not the complexity of the work, nor the lack of resources, nor even a mistake that had to be redone, but simply a dubious assumption underlying a computer algorithm that was written decades ago, but whose implication few project management people understand.)

As shown in Figure 7.28, Activity C has two predecessors: Activity A is an FS predecessor and Activity B is an FF3 predecessor. Activity C is also an SS2 predecessor of Activity D. Because of the FF3 relationship with Activity B, C cannot finish until at least 3 days after B finishes, or Day 18 at the earliest.

C's early finish certainly can't be earlier than Day 18, but based on just the FS relationship with Activity A, C could start as early as Day 6. But the original PDM algorithm was written (and almost every other software package has copied it to the present day) using the assumption that an activity cannot be interrupted once it has started. Thus the software computes C's early start by counting backward from C's early finish, making C's early start Day 14 instead of Day 6. And since the early start of Activity D is an SS2 successor of C, it, therefore, cannot start until Day 16 instead of Day 8. A project manager who is aware of this issue and sees that it is delaying project completion will almost certainly figure out a way for Activity C to accomplish its first two days of work on Days 6 and 7 so that D can start on Day 8 and the entire project will finish on Day 22. However, most project managers are blissfully unaware of this anomaly and so, faced with a schedule that contains 1,500 activities rather than 5, will simply accept the schedule that the software generates, one that might be several weeks longer than it needs to be. The consequence might be significantly reduced EMV or even, yes, people dying unnecessarily.

An interesting twist is to ask the question: How much drag does Activity C have? Notice what would happen to the project duration if C

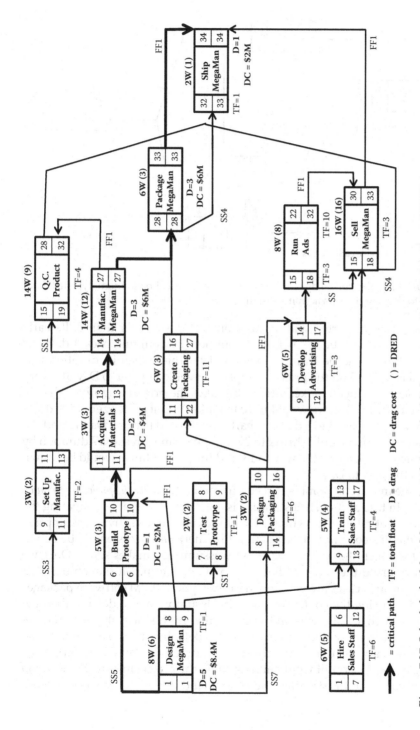

Figure 7.27 Schedule for MegaMan project showing doubled resource estimated durations (DREDs).

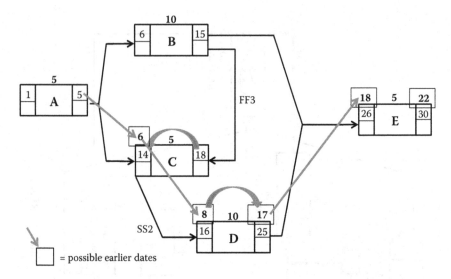

Figure 7.28 Network diagram showing reverse critical path anomaly; later dates due to the "continuous activity" assumption.

were actually shorter, with a duration of 4 days rather than 5. Its early finish would still be set as Day 18, but now by counting back 4 days, its early start would be Day 15—one day later than when its duration was one day more. And the project completion would slip out to Day 31. Every day *shorter* that you make Activity C makes the project *longer* by 1 day. To shorten the project, we would have to make Activity C longer. If C's duration were increased to 6 days, its early start would become Day 13 and the project duration would shrink to 29 days. We can increase C's duration by as much as 8 days, all the way to 13, and the project finish would be drawn in all the way to Day 22.

The important point is to recognize *what is making the project how much longer!* In this particular instance, it is not Activity C that is elongating the project, but rather the combination of the continuous activity assumption and the FF3 constraint from Activity B! There *may* occasionally be activities for which the continuous activity assumption makes sense. Once we start pouring the concrete floor, we may have to finish the job before it sets. But such cases are very much the exception. If PM software packages insist on using the continuous activity assumption as a default, then the drag in such situations is on the constraint: in this case, the FF3 relationship from Activity B to Activity C that is delaying the project completion by 8 days and therefore has critical path drag of 8 days. Computing the drag of the constraint would at least draw the project manager's attention to the fact that it is delaying the project (and undoubtedly has drag cost, too).

Two final points of note:

1. The drag of a constraint is *not* limited to the amount of its lag. If the relationship above were a simple FF with no lag, it would still delay project completion by 5 days.
2. Both FF and SF constraints can cause the reverse critical path anomaly if they immediately precede an SS or SF activity on the critical path.

Summary of the benefits of CPM

Let's summarize the benefits that this wonderful, yet woefully underused technique called CPM offers:

1. **It allows accurate calculation of the project duration, based on the activity duration estimates.** The project may take longer, but will almost never take less time than the CPM estimate. If the project is on a tight deadline, you cannot know that you can make it unless you put together a CPM schedule that meets that deadline.
2. **It provides data for optimization of the schedule.** Not by guesswork and wishful thinking, but by logic and decision making.
3. **It determines the schedule for each activity.** This is extremely important. If we know when each activity must occur, we are in a position to start lining up resources, seeing if they will be available in the required amounts when they are needed. If they aren't, we can see if we can do something about it. Perhaps workers can be hired and trained in advance, or perhaps they can't, so that we will have to delay the activity in question until the resources become available. Maybe that delay will cause the project to lose so much value that it will no longer be worthwhile, in which case, we should cancel the project now instead of wasting resources and money on it for several months before realizing that the investment is doomed to be a bad one.
4. **It provides a "musical score" for the project manager, each activity manager, and each individual worker, showing when and how quickly each must perform.** In addition to the scheduling of resources, this can save valuable time on handoffs between activities. While this may seem a commonplace benefit, delays on handoffs are often responsible for wasting huge amounts of time on complex projects and processes. In organizations where resources are from different functional departments and multitasked, handoffs are particularly inefficient. Project durations can easily be doubled due to the time wasted on handoffs. I strongly recommend that the project schedule always be maintained in hard copy as a network diagram chart on the wall, with each individual responsible for:

  a. initialing and dating the activity box at the time of starting;
  b. informing the project manager and all successor activity managers immediately upon completion (or reaching an SS handoff point); and
  c. initialing and dating the activity box when the activity is completed.
5. **It shows opportunities for savings on resources and/or cost by cutting resources on noncritical path activities** (i.e., trading total float for resources). This, of course, is exactly how the process of resource leveling works, which we will discuss in Chapter 9. In scheduling resources, priority is given to critical path activities, and noncritical path activities are often delayed within their float.
6. **It provides an early warning system once the work has begun on the project.** If the first activity in a six-month project takes four weeks rather than the scheduled three, the dependency relationships will immediately show that, under the current plan, each successive critical path activity will be delayed and the project will finish a week later (at a minimum) than planned. Without the CPM links, one activity slipping has no explicit impact on the rest of the project.
7. **When slippage is detected, CPM provides a tool for seeing how to get back on schedule by fast tracking or "crashing the critical path"** (i.e., increasing resources to shorten the duration of critical path activities). Without an identified critical path, when the project slips there may be no clear reason to target one activity or path instead of four other "workaround" candidates. Additionally, the concept of critical path drag helps make this targeting process much more precise.
8. **With a computer, CPM allows analysis of the project plan through "what-if" scenarios.** Because of the network logic dependencies, changes can be input to the plan as "trial balloons" and their impacts assessed. When slippage or other unforeseen occurrences threaten the plan, a variety of remedies can be tested through CPM modeling, and the optimal remedy can be selected. Without CPM, the impact of such what-ifs is virtually nonexistent because there are no precedents to show the schedule interdependence of activities.
9. **Developing the CPM schedule allows for distinction between project delays that are due either to the nature of the work or the logical order in which it must be performed versus delays that are caused by the lack of sufficient resources.** This is a benefit that is rarely attributed to using CPM; yet it's the only way to achieve anything like right-sized staffing levels in a project-driven organization. Resource delays are something that the organization can usually do something about, given enough time. If the lack of sufficient programmers, jackhammers, or stainless steel is costing the

organization millions of dollars due to project delays, that is something that can be addressed, but only if those delays can be separated out from other, often unavoidable, delays and their impact on the EMVs of all the projects quantified. This process is greatly facilitated by adopting a standard operating procedure that all activity estimates should be based on the underlying assumption of at least one dedicated unit of each resource throughout an activity's performance whenever part-time resources would lengthen the duration.

There may be some corporate projects that would not benefit from CPM scheduling. Perhaps small, simple projects with little corporate investment may be scheduled and performed adequately without CPM techniques. Without a doubt, any corporate project requiring more than 500 person hours, or of more than two months duration, should utilize CPM techniques if it is not to fritter away a large quantity of dollars through avoidable inefficiencies.

Yet corporations continue to shut their eyes to this fundamental and decades-old project management technique. Business schools don't teach it adequately, project managers don't use it competently, and senior managers neither mandate it nor provide the necessary supporting infrastructure of procedures and software. Corporations don't use it because they understand neither its functions nor its value. But time is money, and those project-driven companies that *do* come to understand it and use it as standard practice will sooner or later drive their competition into bankruptcy.

Other methods of scheduling projects

So corporations and government agencies often don't use CPM to schedule their projects. Yet thousands, if not millions, of projects are performed every day. Although many of them seem to be done willy-nilly, with nothing planned or scheduled in advance, some projects certainly start with a schedule in place. So we must presume that *some* scheduling method is being used. What is it?

Well, the first method of scheduling projects is probably the one that is most commonly used—none. That is to say, nothing so formal and systematic as to deserve the term *method*. Typically, the project manager, either in isolation or during a tour through the involved departments, determines that the programming will be done in June and July, the documentation and testing in July and August, sales training and advertising in September, and the product launch is planned for the first Friday in October. This random selection of dates then is often displayed and distributed on a Gantt chart, perhaps to give the whole thing the correct "project management aroma."

The Gantt chart

The Gantt chart is a perfectly respectable project management tool. In fact, it's one of the most venerable project management techniques still incorporated in project management software. First developed by Henry Lawrence Gantt at the Philadelphia Naval Shipyard in 1908, it was initially used to display transatlantic shipping schedules and later for work schedules. It still does this very well.

- The date ribbon at the top of the chart shows the timing of the work.
- The activity bars allow the viewer to see which work is occurring simultaneously.
- The bars are proportional to the length of the activities, so that at a single glance one can get a sense of which activities are longer and which are shorter.
- And, by using different colors and shadings, two different schedules, such as planned versus actual or resource use versus resource availability, can be compared on the same chart.

But the Gantt chart is a display tool; it is not intended for schedule calculation. The schedule is much more easily calculated on a network logic diagram, such as the ones shown in this chapter, and then translated for display into the Gantt format. However the data may be displayed, they should first be computed based on the techniques and metrics of CPM, for which the network logic diagram is well suited.

Figure 7.29 shows the PDM diagram from Figure 7.9 and then a Gantt chart displaying the schedule for the same project.

Backward scheduling

There is a scheduling methodology that is by far the most common technique for scheduling projects. It's called *backward scheduling*, and chances are that we have all been involved in projects in which this technique was used. Here is the way it works:

Imagine that we have an idea for a new product: a remote-controlled lawnmower. This lawnmower will allow the user to mow the entire yard while lounging in a chaise on the backyard deck. The control panel for the mower is a four-way remote controller that also runs the television, DVD player, and cable box. At the front of the lawnmower is a small video camera that will broadcast to the picture-in-picture function on the television screen. Every Saturday afternoon, you will be able to simultaneously watch the ballgame, cut the grass, and chase the neighbor's cat across the lawn if it shows up on your picture-in-picture. What a great product.

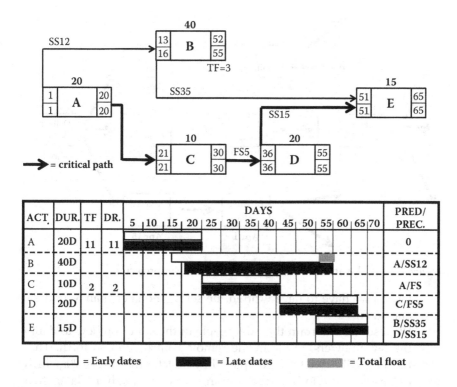

Figure 7.29 Schedule for the new product project in both network diagram and Gantt chart formats.

However, there are schedule issues. Today is October 1 and our TPC Business Case tells us that if we are going to realize the $50 million in sales we are anticipating, we need to have our mowers in the stores by no later than May 1. Every week later will reduce our sales by $10 million, down to zero after the first week in June. How do we schedule our project so that we meet our target date?

We have two all-important bits of scheduling information: today is October 1 and we don't want to go beyond May 1.

- All the mowers have to be in the stores by May 1, which means they have to be shipped out of our manufacturing plant no later than April 24. This means they will have to be packaged no later than April 20, which means we have to finish manufacturing no later than April 10.
- If we are going to sell these lawnmowers, we have to advertise them. We should probably start running the television ads April 15, which means we have to have them produced no later than April 1, which

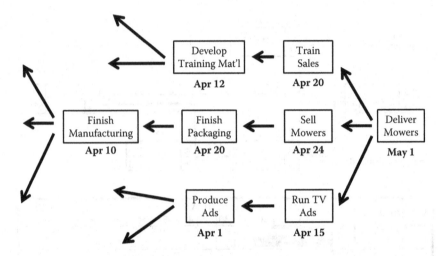

Figure 7.30 Backward scheduling for the lawnmower project.

means we have to have written the scripts no later than March 25, and so on.

- We also need to train the salespeople in the stores on how to demo the lawnmowers, program the remote control, and so forth. That training will have to be completed by no later than April 20, which means we have to have the training materials ready no later than April 12.

Have you seen this sort of schedule before? How many times? Eventually, what we wind up with is a project schedule that looks like the one shown in Figure 7.30.

There is one absolutely astonishing aspect to this method of scheduling: it is simply amazing how often, when all the work that needs to be done is scheduled, and all the "must be done bys" have been accounted for, the schedule brings us right back to—yes, you guessed it—October 1. Simply amazing the way that happens.

Let us consider just how we have developed this schedule. Ignoring for the moment the fact that each duration estimate was generated on the basis of the necessity of meeting the launch date, we also have scheduled each and every handoff on the basis of the date when it has to be done. However, throughout this chapter we have periodically used an algorithm to develop a schedule on precisely that basis—the backward pass of the critical path method.

The key question is: What is the difference between the forward pass and the backward pass of a CPM schedule? Difference, remember, is represented arithmetically by the minus sign (–). The backward pass computes, among other things, the late finish (LF) of each activity; the forward pass computes the early finish (EF) of each activity. The difference (–) between the late finish and the early finish (LF – EF) equals total float. If we do backward scheduling, which is essentially using the backward pass, we have factored out of our schedule all the float, and, thus everything is on the critical path. If any activity slips so much as a day, no matter how trivial its work may be, our project will be running late.

What this means is that the schedule for the lawnmower project is 100-percent critical, and if anything slips, either:

- we will miss our May 1 launch;
- we will have to cut scope, which is adding value (there goes that little camera on the front of the lawnmower);
- we will devastate our resources on exhausting and expensive work-arounds; or
- Any two or all three.

When in one of my executive seminar, I point out the aforementioned problems with the backward scheduling method, someone is almost sure to say: "But you don't understand. In your lawnmower project, it's probably not going to take all the way up to April 20 to do the packaging. It'll probably finish by April 16. The rest of the time is just in there as bank time."

Or safety time or contingency time. It's real name, however, is padding. The problem with padding is the following:

- When padding is built into activity estimates, profitable projects are vetoed, bids are inflated, potential contracts are lost, unnecessary resources are hired, competing projects are delayed, and businesses are bankrupted.
- There is also Parkinson's law to remember. If the deadline for packaging is April 20, you really think it's going to get done by April 16, even if it could?
- But, let us assume that a miracle happens—for our lawnmower project, Parkinson's law is repealed. Manufacturing, scheduled to finish on April 10, actually finishes April 4. Whenever an activity on the critical path (and even elsewhere in the project) finishes early, we would like to start its successors immediately after. However, that never happens in a backward scheduled project. It's hard enough to make it happen in a CPM-scheduled project, where activity managers are usually more aware of the issues of float and the potential of schedules to vary from plan. But in a backward scheduled project, those dates are in concrete; it may be easier to blow up the deliverable than to move a date forward.

So how should the lawnmower project be scheduled? Remember, we want to finish it by May 1, no matter what.

For the moment, let's forget the impending due date. The last thing we need is wishful thinking interfering with our business judgment. If we can't get this project finished by the deadline, we need to know that now, not after we have spent a couple of million dollars in a futile effort to reach an unattainable goal. We need to know what work we have to do and how long it's going to take to do each item. Once we have determined that, then we can start to think about the schedule.

In order to find out what all the work is that needs to be completed, we must do several things.

1. First, we should develop a work scope document listing the specific deliverables, along with appearance and performance standards.
2. Next we should construct a WBS detailing all the work items that have to be completed in order to produce and market those deliverables. This is also a good place to develop our value breakdown structure (VBS), distinguishing between mandatory and optional work and with an estimate of the value-added of each optional item. This could be invaluable if we have to prune scope to meet the launch date.
3. Duration estimates and required resources for each of the activities should be input to the activity level of the work breakdown structure, and all the above should be done without any consideration of scheduling issues.
4. Then the forward pass should be performed. What if it turns out that we get a project completion date of May 29? Isn't missing that May 1 launch date just what we had feared? Should we just scrap the project now? Not at all. We don't have a critical path to optimize as of yet.
5. Now we do the backward pass. This gives us the critical path with drag as well as activities that have float. We now attack the critical path, initially targeting the biggest drag items, to pull in the schedule. We crash by adding resources, we fast track by doing more work in parallel, and, if necessary, we prune scope that is adding less value than the true cost of its budget plus drag cost. When the longest path has been compressed so much that it now has float and another path has become critical, we approach the new critical path in the same way, and we repeat this process until we feel that we have a satisfactory schedule.
6. If at the end of that systematic process, the duration is still such that the TPC Business Case tells us that the project would be unprofitable (i.e., has an unsatisfactory DIPP (Devaux's Index of Project Performance)), then we cancel it now before incurring expense.

But much of the time the above techniques will work to bring the completion date back sufficiently that we can even build in schedule reserve at the end of the project to help ensure meeting the May 1 launch date. Note that this is not the padding we talked about earlier, built as an invisible slush fund into each activity estimate, but a true safety net at the end of the project where it's available if *anything* slips, and activities are likely to start slipping as soon as we try to assign resources and discover they aren't there when we need them.

The program evaluation and review technique (PERT)

Backward scheduling is not an acceptable technique for scheduling projects, and would not be recommended by anyone who is knowledgeable in project management theory. On the other hand, PERT most definitely is recommended by many knowledgeable individuals.

PERT was developed in 1958 by consultants from Booz Allen Hamilton working on the U.S. Navy's Polaris missile project. This was one year after the principles of CPM were set down in construction and plant maintenance industries at the DuPont Corporation. Of course, there is a substantial difference between construction and guided missile development.

Construction and plant maintenance have been performed for centuries. However, in 1958, no one had ever built a Polaris-type missile system that could be fired from a submerged submarine and hit a target a thousand miles away. So, whereas a project manager could get a fairly reliable estimate of how long it would take to build the foundation, erect a wall, or replace a piece of machinery, it was extremely difficult to get similar estimates for how long it would take to develop the Polaris guidance system, the firing mechanism, or the trigger for arming the warhead.

As a result, the scheduling people on the Polaris project developed a formula for predicting the effort and duration of an activity. This formula is based on the normal distribution curve of eighteenth-century French mathematician Abraham de Moivre. In a nutshell, de Moivre's work demonstrated that the results of a set of random trials will distribute themselves around their average value in a bell-shaped curve. The further work of Frederick Gauss analyzed a number of features of what came to be known as *Gaussian curves*. All of this has become a feature of modern-day probabilistic project estimating using Monte Carlo simulation systems that randomly input estimates, run them 5,000 or 10,000 times based on user-defined distributions, and see what the probabilities are for various results.

However, the Booz Allen Hamilton consultants didn't have Monte Carlo systems; they were just trying to get estimates from engineers who were doing work that they had never done before and were reluctant to commit themselves to any one number. So the consultants decided to ask the engineers for not one estimate but three: what estimate is most likely to be correct, what is the most optimistic estimate, and what is the most pessimistic estimate? Then, working backward, they sought to force the estimates into a sort of Gaussian distribution curve by using the formula:

Budget/Duration = (Optimistic + 4 × [Most Likely] + Pessimistic) ÷ 6

This formula would provide a budget/duration weighted toward the most likely estimate, and, in most cases, the most likely estimate will be closer to the pessimistic estimate than to the optimistic. The fastest you can do something is almost always more constrained than how long it will take if everything that can go wrong does.

Let us take an example from the MegaMan development project. Suppose that we ask for the three duration estimates for the *Design MegaMan* activity. The three estimates include:

Most Likely = 8 weeks
Optimistic = 6 weeks
Pessimistic = 13 weeks

According to the PERT formula, the duration for the *Design MegaMan* activity would be

(13W + 4 × [8W] + 6W) ÷ 6
= 51W ÷ 6
= 8.5 weeks

That estimate of 8.5 weeks provides a "quick and dirty" sanity check. If a project manager gets an estimate from an activity manager that seems unrealistic, asking for a most likely, optimistic, and pessimistic estimate can provide insight into whether the single number is way off.

Monte Carlo risk simulations

The PERT methodology above is relatively simple to use for resource usage and/or cost, but schedule is more complicated. The fact that some activities are on the critical path means that the volatility of those specific activities, as opposed to the set of all the project's activities, can make a big difference. We really need to be able to attach the three estimates to each activity's criticality or amount of float in order for the output data to be helpful.

To this end, there are many software packages (which usually market themselves as "risk software") that work with the actual critical path network, only using the three estimates on the basis of user-defined probability distributions to estimate how long the entire project is likely to take. Many of the users of such software accept the output as almost a magically accurate estimate. Unfortunately, it is a classic case of "garbage in, gospel out."

The trouble is that most people think a Monte Carlo system offers some kind of simple panacea, and they don't. It requires a great deal of input effort and understanding to get valuable output. Here are three significant problems that undermine the accuracy of Monte Carlo scheduling systems:

1. What do the terms *optimistic, pessimistic,* and *most likely* (or even *minimum* and *maximum*) really mean? Because they are very different for different people, when one estimator says pessimistic, she is thinking the technology might prove a bit more difficult than expected and thus take 10 percent longer. When her officemate in the next cubicle says pessimistic, he is assuming that his work is interrupted by simultaneous earthquake, tsunami, and nuclear meltdown ("It has happened in the past."). This process can be made a little less subjective if the organization has a robust historical database that permits finding and identifying the smallest, largest, and median durations of similar activities in the past. But the trouble is, what precisely is *similar?* This term can permit a great deal of subjective "cherry picking" of estimates.

2. Most Monte Carlo schedule systems offer a large menu (usually up to 30) of user-selected distribution shapes that can be implemented for each and every activity. But, to select a specific shape for each of 1,500 or more activities (on top of generating the three estimates for each) requires many hours of invested time. In fact, the vast majority of users run the system using one of the default distributions: either the triangular (the three estimates) or the PERT (using the formula to weight the "most likely" estimate and create almost a beta distribution). However, different activity distribution shapes can substantially alter forecast duration. For example, if you run the simulation using the PERT distribution default, you will get an 80-percent confidence duration that is 8 to 15 percent shorter than if you use the triangular distribution. Ain't computers great? Just switch the default and you will shorten your schedule and avoid all those potential liquidated damages. (Sure you will.)

3. If your schedule includes time-based lags, the vast majority of Monte Carlo software will *not* vary the lags. What this means is that if Task A is an SS5 predecessor of Task B, and Task A's three estimates are

6, 12, and 25, the software will set the lag at 5 whether it's assuming the duration is 6, 12, or 25, and that certainly makes no sense and can cause a major distortion in the estimate.

The above are flaws of which most risk software users are bliss-fully unaware. I attended a presentation at the 2009 PMI College of Scheduling by someone who is considered one of the international "gurus" of scheduling. He talked about the great results he got by using a Monte Carlo system for his estimates. I asked him if he didn't find the effort of choosing a distribution shape for each of the thousands of activities enormously laborious. He gave me the reply I expected: "Oh, I run it on one of the default distributions." Which one, I asked. "Oh, I think the triangular, but it doesn't make much difference." Not much difference? Eight- to 12-percent change in duration estimate is "not much difference"? He had no idea because he had never bothered to check out the mathematics. Remember, he is regarded as a scheduling authority. What is the drag cost of an extra 8 to 12 percent on one of his project durations, I wonder?

chapter eight

Activity-based resource assignments

At the end of the critical path method (CPM) scheduling process, we have a schedule for the MegaMan Development Project delivery that shows it will last 34 weeks. We know from the Total Project Control (TPC) Business Case's metrics regarding the value/cost of time that with such a duration we can expect sales of only $2 million, or 20 percent of what they would be if we could finish four weeks earlier.

We also know that our current schedule assumes no slippage. But, how realistic is that? Right now, we have no idea if the resources that we need to meet this schedule will be available when they are needed, and, without that information, we are still stumbling around in the dark. Even our current 34-week duration is probably unattainable.

We know the drag totals of each critical path activity, and have DRED (doubled resource estimated duration) estimates of each activity that might save us a considerable amount of time. But can we implement those estimates? Are these *additional* resources available? Again, we have to do what so many project-driven companies make no attempt to do—systematically determine how the availability of resources might impact our project schedule.

Without having put together the CPM schedule, we would have no way of assessing the impact of resource availability, because we would have only the vaguest idea of when the resources would be needed for each activity. However, with those data available from the CPM schedule, we should not only be able to determine where we don't have sufficient resources, but also the impact that such shortages will have in delaying the schedule.

However, even if we have put together a CPM schedule, we may still not have sufficient information to identify our resource shortages. For that, we also need to have information about resource availability. And in all my years of project consulting, no deficiency has been more striking than the lack of such information. Organizations whose revenues are 100 percent project-driven, and in which every employee is working on four and five projects a week, nevertheless make no attempt to either forecast or track their resource availability and usage. Of course, that means that they can never measure the impact of resource shortages on their project durations, so they can never justify additional resources. Which means greater resource insufficiency, more bottlenecks, more multitasking, and longer projects. What we have here is the paradigm of a vicious cycle.

Let me state, right up front, that it is an absolute requirement for any organization doing multiple projects that it must assemble and maintain an up-to-date resource availability database (usually called a *resource library*). This is mandatory even for those organizations that assemble dedicated project teams. However, for the vast number of companies that operate within a "loose matrix" structure, it is even more important—*it has to be done.*

Unfortunately, it is precisely in such organizations, with shared resources, multitasking, and strong lines of functional management, that maintaining a resource library that reflects assignments and availability across time is the most difficult. It is here that huge inefficiencies and waste are generated. The greatest need for such data lies with the project and project manager, but the means to maintain such a database lies with the functional and senior managers. All too often, these individuals see no personal or organizational benefit in assembling such a database.

One of most important practical benefits of the TPC approach is to show how assigning resources to project activities can benefit, not just the organization, not just the project manager, but also the functional manager. It does so by justifying, in the clearest possible terms, the additional resources that the functional manager *already knows are needed,* and has probably been screaming about for months.

This whole issue raises what I believe is often the case in the project management process: Whenever there appears to be a conflict between the interests of the project and the interests of an individual within the project, the conflict is not real, but rather is caused by process disintegration and/or a lack of understanding of the overall process. In the case in point, if the organization's life blood flows from projects, then all the individuals within it also should benefit by determining how best to do their part in the project management process.

Activity-based resource assignments (ABRA)

The first step in resource scheduling is to determine what resources will be needed for each activity. This is called *activity-based resource assignments* (ABRA). It is the foundation for an important technique in cost accounting that has received much attention in recent years: activity-based costing (ABC). The fundamental concept behind ABC is that the cost for doing work must be assembled and tracked at the level of each activity. ABC is, in its essence, a project management technique. The data for ABC are generated by assigning resources activity by activity, which is ABRA without the "cadabra" of cost. When the dollar rate of the resource usage is included, and computed for each activity, then we have that cadabra, or ABC.

But, even without ABC, ABRA provides indispensable benefits. Resources are the essential items that fuel each activity, and each activity requires its own unique profile of resources. To try to manage a project and its resources at the overall project level is to risk muddying and blurring the work effort. Resources are not interchangeable among all activities; programmers need to work on the programming activities and carpenters on the carpentry activities. Therefore they need to be planned and tracked at the activity level. Yet even where a resource database *is* maintained, it is almost always done at the overall project level, far too high for useful clarity.

In performing ABRA, there are three items of information regarding resource requirements that need to be determined for each activity:

1. **What** resources are needed?
2. **How much** of each resource is needed in order to complete it *within its estimated duration?*
3. **When** will each resource be needed?

In order to keep the size and complexity of our sample project at a manageable level for this book, we have been operating at a fairly high summary level—the 16 activities of the MegaMan Development Project that we identified and scheduled in earlier chapters. We will now assign and schedule our human resource needs (which tend to be the most problematic type of resource) for a few of the activities at this level. However, bear in mind that, on a real project, resource assignments should be made at a much lower level of detail, preferably the activity list (i.e., lowest) level of the work breakdown structure (WBS), and that *all* resources should be assigned and managed in this manner, not just human resources.

In the resource requirements chart in Table 8.1, we have determined how much of each resource would be needed to complete each of these activities in the estimated amount of time. As mentioned earlier, the operating assumption for each activity is dedicated resources throughout the activity's duration, except where it would not be advantageous to do so. For example, *Acquire Materials* is a three-week activity, but most of that time will be spent waiting for delivery. To have the material acquisition clerk and the inventory controller dedicated to this activity for three weeks would be absurd, and affect the duration not a whit. Therefore, the project plan is adjusted so that the clerk works just the first three days of the activity, ordering the materials, and the inventory controller works just the last week, checking in the materials. (Of course, this would probably be a nonissue if we were assigning resources at a lower level of detail, where the ordering and checking in activities would be scheduled as separate activities.)

The required resources having been assigned, the next step is to match them against the resources within the organization. Resources

Table 8.1 Required Resources and Work Days for MegaMan Activities

Activity	Duration	Assigned resources	Work days
Design MegaMan	8W	1 Product Manager	40WD
		2 Product Designer	80WD
		1 Design Engineer	40WD
		1 Drafter	10WD
Build Prototype	5W	2 Design Engineer	50WD
		1 Manufacturing Engineer	10WD
Test Prototype	2W	2 Design Engineer	20WD
		1 Test Engineer	10WD
Design Packaging	3W	1 Packaging Designer	15WD
		1 Packaging Engineer	15WD
		1 Manufacturing Engineer	5WD
Create Packaging	6W	1 Manufacturing Engineer	30WD
		1 Manufacturing Supervisor	30WD
		5 Manufacturing Worker	150WD
Package MegaMan	6W	1 Package Manager	30WD
		1 Packaging Supervisor	30WD
		10 Packagers	300WD
Ship MegaMan	2W	1 Shipping Clerk	4WD
		1 Shipping Supervisor	10WD
		2 Shippers	20WD
Set Up Manufacturing	3W	2 Manufacturing Engineer	30WD
		1 Manufacturing Supervisor	6WD
		4 Manufacturing Workers	60WD
Acquire Materials	3W	1 Material Acquiring Clerk	3WD
		1 Inventory Controller	5WD
Manufacture MegaMan	14W	2 Manufacturing Engineers	140WD
		1 Manufacturing Supervisor	70WD
		20 Manufacturing Workers	1,400WD
Quality Control Product	14W	1 Quality Engineer	28WD
		1 Quality Control Worker	28WD
Develop Ads	6W	1 Marketing Manager	12WD
		1 Ad Writer	30WD
		1 Graphics Specialist	15WD
Runs Ads	8W	1 Marketing Manager	10WD
Hire Sales	6W	1 Recruiter	30WD
		1 Sales Manager	15WD
Train Sales	5W	1 Instruction Designer	25WD
		1 CBT Designer	25WD

(Continued)

Table 8.1 *(Continued)* Required Resources and Work Days for MegaMan Activities

Activity	Duration	Assigned resources	Work days
		1 Trainer	10WD
		5 Telemarketer	25WD
		5 Field Sales	25WD
Sell MegaMan	16W	1 Sales Manager	80WD
		5 Telemarketer	400WD
		5 Field Salespersons	400WD
TOTAL			3,691WD

within an organization are located in the departments that comprise the hierarchical structure of the company. This is also called the organizational breakdown structure, or OBS. You may recall that we displayed the OBS of MegaProdux, Inc. in Chapter 5, Figure 5.1. We show it again in Figure 8.1, but this time with the resources available in each department.

In addition to showing the internal resources of each department in MegaProdux, Inc., the OBS also should contain cost information about each resource. This is where the "C" in ABC comes from. For human resources, these cost data should be based on overhead-burdened salary standards, not on the actual salaries of the individuals. This keeps individual salary information confidential, while providing a project manager and team with approximate numbers for planning cost without the necessity of knowing exactly which individuals will be assigned. For example, it is sufficient to know that the standard cost for a shipping clerk is $500 per week; we don't need to know whether it will be Angus Patel at $475 per week or Jean Smith at $550 per week.

Activity-based costing is a much more complex subject than we can adequately cover here. Readers who are interested in exploring it in depth should read the many fine books on the subject. For our purposes, we need three things:

1. A fairly accurate budget for each resource being used, for each activity, and for the entire project. To this end, standard salaries, burdened by overhead costs (health and dental insurance, clerical support, equipment, energy, and physical plant costs, such as rent, etc.), are adequate. These will allow comparisons and decision making both within a project and across projects.
2. A schedule that shows how the cost of a resource changes over time. For example, each shipper may be available 60 hours per week, but while the first 40 hours may be at a rate of $12, the other 20 hours may be at the overtime rate of $18. This information needs to be

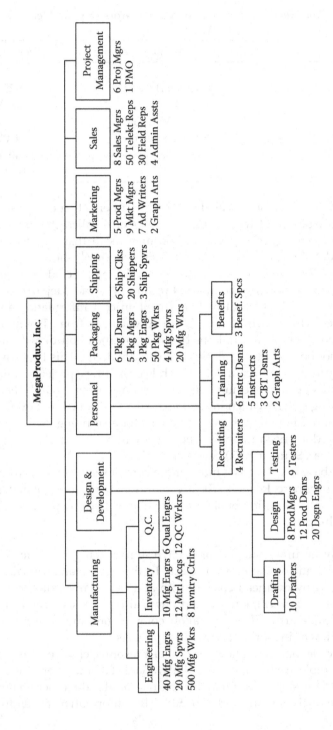

Figure 8.1 OBS of MegaProdux, Inc., showing internal resources.

taken into account both when planning a project and when determining what staffing level to maintain in the shipping department.
3. A schedule that shows how the availability of a resource varies over time. It is this "calendar" information that transforms the OBS into a resource library, the implication of the term *library* being that it allows us to see what resources are available, and which have been "checked out," and when they are expected back, so that we can schedule the resources for our project over time, as we need to do.

If the MegaProdux OBS contains the necessary costing information, we can get our labor budget for

- each activity;
- each summary-level activity;
- each supporting MegaProdux department specifically for the MegaMan project; and
- the entire project. In fact, this is precisely how a project budget should be generated.

Almost all the project management software packages that are currently available incorporate the functionality of a resource database. Some do this better than others. Some allow for the work hours of a resource to change in both cost and availability several times a day while others are less flexible. There may be desired ways of "slicing and dicing" resource usage that are not available in less expensive packages. However, almost all provide the ability to do at least *some* rudimentary ABRA and ABC, and, like all the other project management techniques we have examined, this functionality is largely ignored in most organizations. Why? Because it takes time and effort to keep such a database maintained and up-to-date, and organizations just can't be bothered. After all, they have projects to do, right? And they are underresourced as it is.

They will remain under-resourced, too, because the only way to justify additional resources in a project-driven organization is by contrasting the costs of those resources versus the impact of resource shortages on project schedules and EMV. Without a resource library, this is all but impossible.

Assigning the resources to the MegaMan project

So now let us load the MegaMan project resource requirements into our software and assign the necessary resources from the organizational OBS. As the resources are assigned, the cost data come with them. As shown in Table 8.2, we now have the labor budget for each activity.

Table 8.2 Required Resources, Work Days, and Labor Budget for MegaMan
Project Activities

Activity	Duration	Assigned resources	Work days	Cost	Labor budget
Design MegaMan	8W	1 Product Manager	40WD	$24,000	$80,500
		2 Product Designers	80WD	$40,000	
		1 Design Engineer	40WD	$14,000	
		1 Drafter	10WD	$2,500	
Build Prototype	5W	2 Design Engineers	50WD	$17,500	$21,500
		1 Manufacturing Engineer	10WD	$4,000	
Test Prototype	2W	2 Design Engineers	20WD	$7,000	$10,000
		1 Test Engineer	10WD	$4,000	
Design Packaging	3W	1 Packaging Designer	15WD	$6,000	$13,250
		1 Packaging Engineer	15WD	$5,250	
		1 Manufacturing Engineer	5WD	$2,000	
Create Packaging	6W	1 Manufacturing Engineer	30WD	$12,000	$51,000
		1 Manufacturing Supervisor	30WD	$9,000	
		5 Manufacturing Worker	150 WD	$30,000	
Package MegaMan	6W	1 Package Manager	30WD	$10,500	$63,000
		1 Packaging Supervisor	30WD	$7,500	
		10 Packager	300WD	$45,000	
Ship MegaMan	2W	1 Shipping Clerk	4WD	$600	$6,100
		1 Shipping Supervisor	10WD	$2,500	
		2 Shippers	20WD	$3,000	
Set Up Manufacturing	3W	2 Manufacturing Engineers	30WD	$12,000	$25,800
		1 Manufacturing Supervisor	6WD	$1,800	

(Continued)

Table 8.2 (Continued) Required Resources, Work Days, and Labor Budget for MegaMan Project Activities

Activity	Duration	Assigned resources	Work days	Cost	Labor budget
		4 Manufacturing Workers	60WD	$12,000	
Acquire Materials	3W	1 Material Acquiring Clerk	3WD	$600	$1,600
		1 Inventory Controller	5WD	$1,000	
Manufacture MegaMan	14W	2 Manufacturing Engineers	140WD	$56,000	$357,000
		1 Manufacturing Supervisor	70WD	$21,000	
		20 Manufacturing Worker	1,400 WD	$280,000	
Q.C. Product	14W	1 Quality Engineer	28WD	$8,400	$12,600
		1 Quality Control Worker	28WD	$4,200	
Develop Ads	6W	1 Manufacturing Manager	12WD	$6,000	$20,250
		1 Ad Writer	30WD	$10,500	
		1 Graphics Specialist	15WD	$3,750	
Runs Ads	8W	1 Marketing Manager	10WD	$5,000	$5,000
Hire Sales	6W	1 Recruiter	30WD	$10,500	$16,500
		1 Sales Manager	15WD	$6,000	
Train Sales	5W	1 Instruction Designer	25WD	$8,750	$35,500
		1 CBT Designer	25WD	$8,750	
		1 Trainer	10WD	$3,000	
		5 Telemarketer	25WD	$6,250	
		5 Field Salespersons	25WD	$8,750	
Sell MegaMan	16W	1 Sales Manager	80WD	$32,000	$272,000
		5 Telemarketer	400WD	$100,000	
		5 Field Salespersons	400WD	$140,000	
TOTAL			3,691WD		$349,250

Again, these are just the labor budgets. Accurate costing requires that *all* types of resources (equipment, materials, etc.) be assigned to the activities in order to determine the budget for each activity. To save time on our MegaMan project example, we simply are going to estimate the total costs by assuming the labor budget is a variable percentage, depending on the type of activity. Each type of activity is a different type of work and can be expected to be more or less labor intensive. Manufacturing, for instance, will require more equipment and materials than designing, while advertising will require paying for placing the ads. For each of the four summary-level activities, we will estimate what percentage of the total budget the labor budget represents (Table 8.3). The one exception will be *Run Ads*, where a major expense is incurred in the placement of the advertisements. An additional $400,000 will be added for this.

Project management work and costs

At this point, we must say just a word or two about the Project Management branch in our work breakdown structure. Project management uses resources, costs money, and has specific activities in any project that should be scheduled. Many of these activities are level of effort (LOE) activities, which are repeated activities in support of the rest of the project work: *File Progress Reports Weekly*, or *Visit Subcontractor Site Every 2 Weeks during Manufacturing* might be examples of such LOE activities. LOE activities are in support of the project and should *never* be on the critical path or have drag and drag cost. However, some project management activities, such as *Develop Detailed Specs*, are associated with project artifacts, have a specific start and finish, and may sometimes be on the critical path.

Because we have wanted to keep our sample project as simple as possible (and the reader has probably noticed that, even with only 16 activities, it has nevertheless become a bit complex), we have omitted the detail activities that might appear under the Project Management branch. However, let me emphasize that on a real project we would definitely want to include them, even the LOE activities. The LOE activities are particularly important if our project is being tracked on the basis of *earned value* (which we will discuss shortly) and progress payments are based on total earned value. Project management costs money, and tracking the budgets for work performed can be of great importance in maximizing progress payments.

For our purposes, though, we are simply doing what many companies do for internal projects—treating project management work as though it were overhead on the project and giving it a fixed budget for all work of $100,000.

Table 8.3 Activity and Summary Activity Total Budgets Derived by Estimating Percentage of Labor Costs

Summary activity	Activity	Labor budget	Labor % of total budget	Total budget	Summary budget
Product Design			75%		$149,333
	Design MegaMan	$80,500	75%	$107,333	
	Build Prototype	$21,500	75%	$28,667	
	Test Prototype	$10,000	75%	$13,333	
Manufactured Product			30%		$1,323,333
	Set Up Manufacturing	$25,800	30%	$86,000	
	Acquire Materials	$1,600	30%	$5,333	
	Manufacture MegaMan	$357,000	30%	$1,190,000	
	Quality Control Product	$12,600	30%	$42,000	
Packaging & Distribution			50%		$266,700
	Design Packaging	$13,250	50%	$26,500	
	Create Packaging	$51,000	50%	$102,000	
	Package MegaMan	$63,000	50%	$126,000	
	Ship MegaMan	$6,100	50%	$12,200	
Sales & Marketing Strategic Plan			20%		$2,146,250
	Develop Ads	$20,250	20%	$101,250	
	Runs Ads	$5,000	20% + $400K	$425,000	
	Hire Sales	$16,500	20%	$82,500	
	Train Sales	$35,500	20%	$177,500	
	Sell MegaMan	$272,000	20%	$1,360,000	
Project Management					$100,000

Total budget and starting DIPP

Now, we sum the budget numbers all the way to the top for the entire project (Table 8.4).

So we have got a project with a budget of $3,976,783 that, if it's completed in 30 weeks, has an expected monetary value of $10 million. This would give us the following initial DIPP (Devaux's Index of Project Performance):

DIPP = EMV (expected monetary value) divided by cost ETC (esti-mate-to-complete) = $10,000,000 divided by $3,976,783 = 2.51

However, the current schedule calls for it to take 34 weeks, and, at 34 weeks, the EMV is only $2 million. Thus, the current DIPP is much less than 1:

DIPP = $2,000,000 divided by $3,976,783 = 0.50

We have four alternatives:

1. Cut costs drastically.
2. Reduce the duration.
3. Increase scope to increase expected value (with probable increases in cost as well).
4. Cancel the project.

Analyzing and implementing the DRED

With our activity budgets computed, we can now calculate the estimated DRED cost of the critical path activities, or what the cost *might* be of adding resources in order to reduce the drag. This then can be compared to the dollar value of the reduction in drag costs due to the time that would be saved on that activity.

I use the term *might be* rather than *would be* because adding resources would probably *not* result in a "straight-line" increasing of the cost of the activity for two reasons:

1. When the resources-per-time-unit are increased, the *amount of time* over which they are used decreases. This is a major advantage of utilizing the DRED duration to shorten an activity. If the activity's

Table 8.4 Labor and Total Budget for the MegaMan Development Project

Project name	Labor budget	Total budget
MegaMan Development	$991,600	$3,976,783

duration is halved when the resources-per-week are doubled, then we have a 1:1 ratio, which would leave the resource usage unchanged from the original duration. If the cost rate of that usage also remains the same, the cost would stay the same.

2. It is very possible that we may have to go to time-and-a-half labor rates, or pay some other premium, in order to get the additional resources. This would have to be determined during the resource scheduling and leveling processes.

However, if the resources are readily available, we can compute the cost of the DRED level of resources. We then will have some numbers that, while not necessarily precise, are ballpark figures that we can analyze and, ultimately, check for accuracy when we do resource scheduling against the resource availability database. These will allow us to compare the DRED cost with the drag cost and try to arrive at the most profitable schedule. Notice how analyzing resources on this basis starts us moving toward resource levels that are "right-sized" in terms of their value to the project and to the organization.

The DRED is an indication of *resource elasticity*, how much shorter or longer an activity's duration might be if resources were added. The duration estimate based on the assumption of doubled resources is a valid presumption in that the activity manager and/or subject matter expert has said: "Yes, given twice the resources per day, we could probably finish this activity in x number of days." However, it would be dangerous to assume that the DRED represents some sort of precision metric or constant, based on "straight-lined" logic: If doubling the resources cuts the project duration by 20 percent, then tripling the resources should cut it by 40 percent, or 30 percent, or any other number. Such an inference would always be very dependent on the exact nature of the work. The Law of Diminishing Returns may determine that, on a given activity, there is zero additional impact, or even a *negative* impact, from tripling resources.

Rather, what the DRED does is give the project manager a sense of what *might* be the effect of adding resources to an activity. For instance, if an activity has a duration estimate of 12 weeks and a DRED of 10 weeks, it *may* be that adding just 50-percent more resources, rather than doubling, we *might* succeed in getting an 11-week duration, and it *may* be that tripling the resources would reduce the duration to 8 weeks. But the project manager cannot assume that. Such an inference must always be run past the original estimator. "Okay, Joe, you said that if you had twice the resources, you could shave the time needed for this work by two weeks. But this activity is on the critical path with drag of three weeks. What would it take to cut it by three weeks? Is it possible? Would putting six people on it, instead of the current two, be enough? Would two do the trick?" If Joe says that even if you get the entire U.S. Marine Corps working on it, you are not going to be able to cut it by three weeks, then that is the reality with which the project plan must be reconciled.

Ultimately, we would have to find out whether such resources (and, remember, the assumption is that they are identical to the original resource estimates: equal skills, equal training, etc.) would be available. The cost of the additional resources as well would have to be computed and weighed. But, for the moment, we just want to know whether or not, *if* the additional resources were applied, they would have a beneficial impact. To just assume that they are unavailable and, therefore, ignore what impact they *could* have is to turn a blind eye to the fact that

- resources that may be completely unavailable at current market prices might become quite plentiful when the cost of *not* having those resources (i.e., drag cost) is quantified and used as justification for paying more; and
- if the potential beneficial impact of increased resource levels is ignored in project after project throughout the organization, there will never be justification for those additional resources.

The DRED comes with a cost component: What is the additional cost, without overtime or other premium, when we double the resources? It's not as simple as doubling the budget, because the amount of time during which we use *all* the resources, or the work hours, will be cut by the difference between the original duration estimate and the DRED. That, after all, is why we would be implementing the DRED. The tentative *DRED Budget* (prior to adjustments for availability premiums) could be computed as follows:

DRED Budget = 2 × (Activity Budget) × (DRED ÷ Duration)
Let us take the activity *Develop Advertising* in Table 8.5.
Based on this, the *Develop Ads* DRED Budget would be

2 × ($101,250) × (5 divided by 6)
= $202,500 × 0.83
= $168,075

This, in turn, leads to *DRED Cost*, or the cost of implementing the DRED duration, as being equal to

Table 8.5 Duration, DRED, and Budget for Develop Ads Activity

Activity	Duration	DRED	Budget
Develop Ads	6W	5W	$101,250

DRED Budget – Activity Budget
= $168,075 – $101,250
= $66,825

In other words, doubling the resources on *Develop Ads* will increase the cost of resources by $66,825, before taking into account overtime or other availability premiums.

In this case, *Develop Ads* is off the critical path and, therefore, has no associated drag cost. However, if it were *on* the critical path, the one week saved, at $2 million, would be easily worth the cost. In fact, the premium would have to be very large (or the drag cost a lot smaller) for it *not* to be worthwhile to implement the DRED.

The chart in Table 8.6 shows the results of this analysis for the critical path of the MegaMan Development Project. It shows the cost of implementing each DRED duration, and then it shows the results, in schedule, in difference in each activity's true cost, and in the project's expected project profit. Notice that the activities are arranged in descending order of the benefit of utilizing the DRED.

The numbers in the DRED chart are based on a hypothetical case study, yet they are fairly typical in terms of the economic situation they portray; the cost, without availability premiums, of the resources needed to shorten a project is almost invariably less *by huge amounts* than the cost, in EMV reduction, of *not* shortening the project. This suggests that the availability premiums that are justifiable are also huge. For example, the net gain of $6 million on the *Package MegaMan* activity suggests to me that we could pay out a bonus of $100,000 for each of 20 packagers working on the activity at its DRED level of resourcing, along with a cool $250,000 for the two supervisors and two managers, and *still* make $3 million more in profit than if we just sit on our thumbs and say, "Gee, we don't have any more packaging personnel." Think we could find 10 more packagers for a $100,000 bonus for three weeks work? You know what? I bet we could get them for a $2,000 bonus, and increase our profit by almost the full $6 million. (Of course, if we don't want to anger the workers originally assigned, we should let them share in the bonus, too.)

So why are such premiums not routinely implemented on projects? Because the vast majority of business organizations aren't even using the rudiments of traditional project management, much less these new TPC metrics. They aren't even using the critical path method, so how can they ever determine where they need more resources? The biggest difficulties are in determining *where* the additional resources will do *any* good and the *most* good.

This is precisely the analysis that the TPC metrics drag, drag cost, DRED, and DRED cost are designed to facilitate. Without that analysis, paying out availability premiums is usually just throwing money

Table 8.6 Cost/Benefit Analysis of Using DREDs on the Critical Path Activities

Activity	Duration	Budget	Drag	Drag cost	DRED	DRED cost	Time saved	New drag cost	Net $ gained
Package MegaMan	6W	$126,000	3W	$6.00M	3W	$0	3W	$0	$6.00M
Design MegaMan	8W	$107,333	5W	$8.40M	6W	$53,662	2W	$4.40M	$3.95M
Manufac. MegaMan	14W	$1,190,000	3W	$6.00M	12W	$850,000	2W	$2.00M	$3.15M
Build Prototype	5W	$28,667	1W	$2.00M	3W	$5,733	1W	$0	$2.00M
Ship MegaMan	2W	$12,200	1W	$2.00M	1W	$0	1W	$0	$2.00M
Acquire Materials	3W	$5,333	2W	$4.00M	3W	$5,333	0W	$4.00M	$0

away, so you are probably better off not paying them at all. Of course, this doesn't stop some organizations from paying such premiums, anyway. How often do companies pay early or on-time delivery incentives to suppliers, and then have the new parts sit around for weeks until the rest of the project team is ready to use them? That couldn't ever happen, could it?

Making the scheduling decisions

Many people find it easier to make decisions when there are no data available. This is because, without data, you can never be *shown* to be wrong. But, you can still be wrong, even if the data don't immediately show it, and, unfortunately, when the data do eventually come to light, it may be under the heading Corporate Bankruptcies.

The TPC methodology does *not* make the decisions for the project manager. It simply provides data on which project decisions can be based. Many decisions that previously might have been seen as close issues, and where the wrong call may well have been made, should now become no-brainers. But, for other decisions, the data may merely highlight the evenness of the competing solutions.

The data from the above chart relating the DRED to potential gain now needs to be applied to the actual details of the project schedule. We must return to the network logic diagram of our MegaMan Development Project to make each decision and change sequentially, assess the impact of each, and then move to the next decision (Figure 8.2).

From the previous chart, we could see that *Design MegaMan*, with a budget of $107,333, is costing us $8.4 million in drag. Going to the DRED would cut the drag by two weeks (and the project duration to 32 weeks), saving us $4 million in drag cost for an additional $53,662 in resources, a net gain of $3.95 million. Clearly, the DRED should be implemented if at all possible. However, why not keep adding resources, if we can? And, at what point should we stop? We know that the activity is quite resource-elastic, because doubling the resources is expected to reduce the duration by two weeks, or 25 percent.

In addition, *Manufacture MegaMan*, *Package MegaMan*, and *Ship MegaMan* all are on the critical path, with drag, and have potential DRED gains that make them attractive targets for additional resources. By contrast, notice that *Acquire Materials* is not at all resource elastic. Its two weeks of drag cannot be removed by doubling the resources (although, of course, it may be possible to shorten the activity by changing resources— new suppliers, especially with additional dollars going to early-delivery incentives, might do wonders).

It's time for a conference with the activity managers of all the resource-elastic critical path activities, and, if necessary, their subject matter experts. The project manager should start by displaying the current

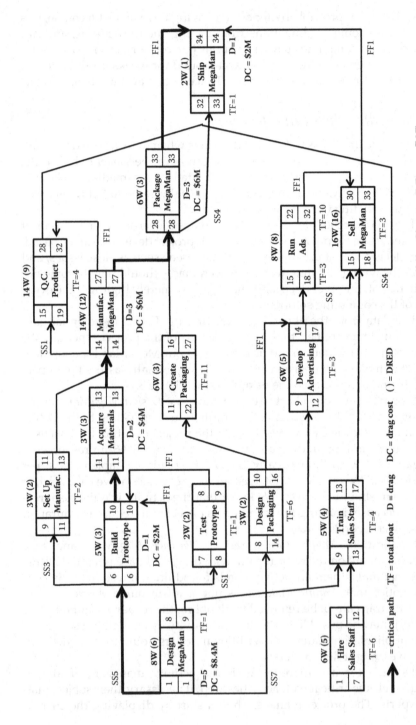

Figure 8.2 Current CPM schedule for the MegaMan project, showing drag totals (from Chapter 7, Figure 7.27).

network schedule, showing the scheduling and EMV implications, and explaining how the DRED estimates of these particular activities offer possibilities for improving the current scenario.

In looking at the *Design MegaMan* activity, we see that all the drag is in the first five weeks, with the work that has to be performed in order for the next critical path activity, *Build Prototype*, to start. We might be able to use the DRED to save two weeks, thereby shortening the start-to-start lag value to three weeks. Is this feasible? The activity manager for *Design MegaMan* immediately says no; he doesn't have any additional resources available to put on this activity. Immediately the project manager stops him.

"You don't understand," she says. "Leave me to worry about the resources. What I'm asking is, if we *were* able to double the resources on this activity, would we be able to shorten it by two weeks and shorten the SS lag to three weeks? And, if we got you even more resources—never mind from where—would you be able to knock any more time off it?"

One interesting thing about this approach is that the activity and functional managers are often so delighted at the prospect of hiring additional resources that they will eagerly collaborate in the project optimization process. It may be important for the project manager to stress that the additional resources can only be purchased through the confident promise of the forecasted time savings. It would be most unfortunate if additional resources were hired at exorbitant expense, only to discover that the activity's drag is left unchanged.

After more careful thought, the activity manager replies that bringing on a bunch of additional people to the *Design MegaMan* activity would only muddle things (the Too Many Cooks syndrome), but that the DRED reduction might be accomplished by using one additional design engineer, and having the entire activity team other than the drafter work 80-hour weeks. There may be a risk, he warns, in that tired brains do not operate as imaginatively, and are more subject to errors. Perhaps, suggests the project manager, a nice project bonus for good work and on-time completion will keep the brains sufficiently fresh. The activity manager narrows his eyes and nods, but now stresses that this really is all he can offer; further reduction is not feasible. The project manager agrees, and the change is implemented.

The reduced duration of six weeks also reduces the SS lag to three weeks and the drag to three weeks. This, in turn, increases the project's EMV by $4 million, to $6 million. Additionally, since the activity duration is now just six weeks, we don't need an SS7 relationship with *Design Packaging*; a simple FS relationship will do. The adjusted network logic diagram is shown in Figure 8.3, with a duration of 32 weeks (correlating to an EMV of $6 million).

With the increased EMV, the DIPP, previously 0.50, now grows to more than 1.0; how much more we won't know until all the resource and

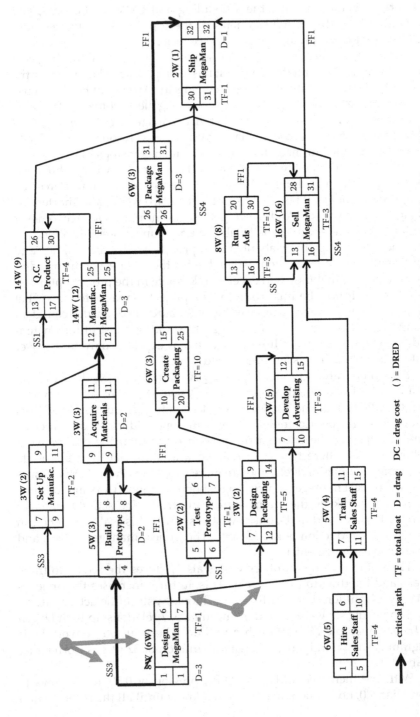

Figure 8.3 Schedule for the MegaMan project after implementing the DRED on the *Design MegaMan* activity.

availability premium decisions have been made. But at least we now have a schedule that suggests the project will be profitable. Yet it's important to recognize that this entire remedial process would be quite impossible in the typical organization where the budget is capped up front. If the project manager's mandate is to spend no more than, for example, $3,980,000, then there would be no room for this kind of maneuvering. We would have had to cancel the project by now. The concept of "spend money to make money" is alien to any environment that doesn't recognize that projects are investments.

Now the conference discussion turns to the other critical path activities where time might be saved. The drag on the other resource-elastic critical path activities *Manufacture MegaMan* and *Package MegaMan* (as well as the inelastic *Acquire Materials*) remains at three weeks. This is particularly fortuitous in the case of *Package MegaMan*, since its DRED shows it to be so resource elastic (1:1 in terms of time gained for doubled resources) as to be potentially doable in three weeks. Is this indeed the case? The activity manager confirms that this is so, although he warns that the number of in-house packagers is limited and that the Human Resources Department has been particularly reluctant to approve the hiring of additional packagers. It seems that the booming economy has made the going rate for these unskilled laborers more than some cost-cutting maven is willing to pay. The project manager nods absently, mutters what sounds like: "Leave that to me," and changes the duration of *Package MegaMan* to three weeks, and its lag relationship with *Ship MegaMan* to SS2. This produces the network logic diagram seen in Figure 8.4.

Now we have a schedule that not only meets the original requirement of allowing our product to reach the retail outlets by the start of the shopping season at the beginning of Week 31, but we have even built in an extra week of schedule reserve. Remember, though, *it is also worth an additional $400,000 if we don't need to use it,* according to the TPC Business Case projections, and our project EMV would rise to $10,400,000.

In looking at the latest network logic diagram, three things should stand out:

1. There are now two separate critical paths, which means two paths of the same length.
2. The drag of almost all the activities (all except the source and the sink activities) on both critical paths is 0. This makes sense when you realize that an activity's drag is equal to the total float of the parallel activity with the least total float. Since the parallel paths are both critical, activities on each will have total float of 0, mutually limiting the drag of the parallel path activities to 0.
3. The total float of the activities that are not on the critical paths also have shrunk to small numbers, in most cases no more than two weeks.

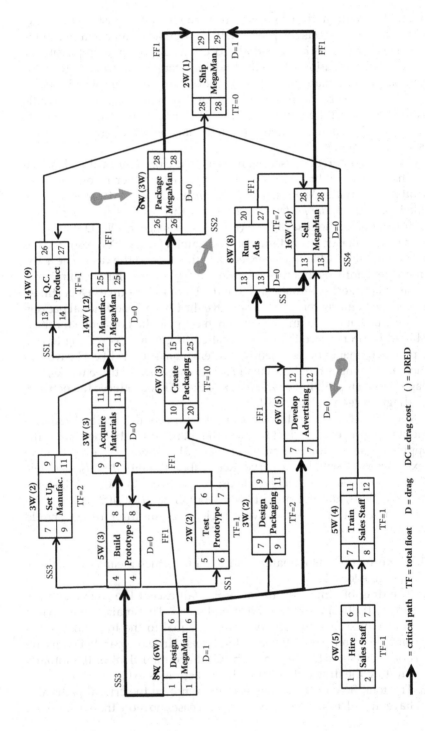

Figure 8.4 Schedule for the MegaMan project after implementing the DRED on the *Package MegaMan* activity.

All this is indicative of a project schedule that is being compressed toward its minimum. Between the start of Week 4 and the end of Week 28, any further compression will require similar duration reductions on the parallel critical path. If a two-week compression *is* accomplished, even more paths will then become critical, making further shortening even more complex and less likely.

The two activities that still incorporate drag are the ones where nothing is parallel: the first three weeks of *Design MegaMan* and the last week of *Ship MegaMan*. We already know from talking with the activity manager for *Design MegaMan* that there is little hope of that activity being compressed further. However, that last week of *Ship MegaMan* seems to offer an opportunity. It has a duration of two weeks and a DRED of one week. Currently, it is on the critical path because of the FF1 relationships with *Package MegaMan* and *Sell MegaMan*.

However, it seems somewhat wasteful to have to take an entire extra week to ship the packaged product. Because of the DRED, we know that all the actual work of shipping could be accomplished in one week. Yet we are adding a week of drag, at a cost of $400,000, for that one week. This requires a discussion with the activity managers of all three activities. The managers of *Package MegaMan* and *Ship MegaMan* are already present, so we send out and ask the activity manager for *Sell MegaMan* to join us.

The ensuing discussion soon makes it clear that the one-week lags have been input as a sort of "safety valve." Just in case there are any problems with any of the three activities, the three activity managers want to make sure there is time to coordinate things and work the kinks out. Which is fine, *except it's going to cost us $400,000*.

When this cost factor is explained to the activity managers, they understand the implications and are only too eager to work around the problem. The activity manager for *Sell MegaMan*, who is also the marketing manager assigned to that activity, agrees to hire an administrative assistant whose job it will be to coordinate the selling, packaging, and shipping. And, in going to the DRED level of resources, the activity manager for *Ship MegaMan* agrees to adopt a three-shift, six-days-a-week schedule that will allow shipping to continue for 32 hours after both selling and packaging are complete. This means that the last shipment will leave the loading dock at midnight Saturday night, still in time to reach the stores by 9 a.m. Monday morning. All resulting in additional EMV of $400,000.

Our schedule now looks as shown in Figure 8.5.

The budgets of a few of the activities will have changed to generate the new durations. The new data are displayed underlined in Table 8.7.

Notice that, even though the DRED was used by the project manager to determine which critical path activities were resource elastic, in only one case was the activity duration reduced by doubling the resources. In the two other instances, once the activities were identified as being

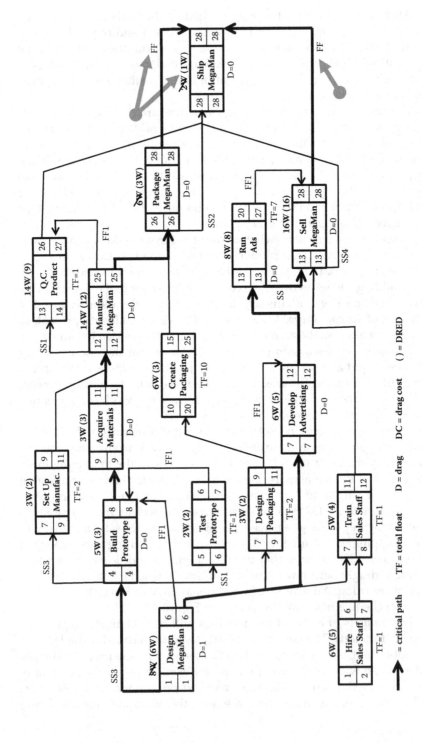

Figure 8.5 Schedule for the MegaMan project after implementing the DRED on the *Ship MegaMan* activity.

Table 8.7 Impact of Applying Additional Resources on the MegaMan Development Budget

Activity	Old duration	New duration	Old res. assignments	New res. assignments	Original words	New words	Original labor budget	New labor budget	Original total budget	New total budget	Budget increase
Design MegaMan	8W	6W			170WD	310WD	$80,500	$140,500	$107,333	$187,333	$80,000
			1 Product Mgr.	1 Product Mgr.	40WD	60WD	$24,000	$36,000			
			2 Product Dsnr.	2 Product Dsnrs.	80WD	120WD	$40,000	$60,000			
			1 Dsgn. Engr.	2 Dsgn. Engr.	40WD	120WD	$14,000	$42,000			
			1 Drafter	1 Drafter	10WD	10WD	$2,500	$2,500			
Package MegaMan	6W	3W			170WD	310WD	$63,000	$63,000	$126,000	$126,000	$0
			1 Package Mgr.	2 Package Mgrs.	30WD	30WD	$10,500	$10,500			
			1 Pkgng. Spvr.	2 Pkgng. Spvr.	30WD	30WD	$7,500	$7,500			
			10 Packagers	20 Packagers	300WD	300WD	$45,000	$45,000			
Sell MegaMan	16W	16W			880WD	960WD	$272,000	$288,000	$1,360,000	$1,440,000	$80,000
			1 Sales Manager	1 Sales Manager	80WD	80WD	$32,000	$32,000			
			5 Telemarketers	5 Telemarketers	400WD	400WD	$100,000	$100,000			
			5 Field Sales	5 Field Sales	400WD	400WD	$140,000	$140,000			
				1 Admin. Asst.		80WD		$16,000			

resource elastic, better and more precise ways were found of applying resources in order to shorten the activity.

Table 8.7 shows that the total budget for *Package MegaMan* was unchanged, while those for *Design MegaMan* and *Sell MegaMan* both increased by (coincidentally) $80,000. The total budget for the project has been increased (so far as we can determine at this stage) by $160,000 to $4,036,783, while the projected duration has gone from 34 weeks to 28 weeks, and the EMV from $2 million to $10.8 million. The DIPP analysis now looks much healthier:

DIPP = $10,800,000 divided by $4,036,783
= 2.68

TPC value scheduling

Earlier, you will recall, we developed a value breakdown structure (VBS) that drove the concept of managing the project's EMV down to the activity level by determining the value that each activity was adding to the project's EMV. The updated VBS is displayed in Figure 8.6, with activity value-addeds translated into percentages of the overall project EMV of, now, $10.8 million.

We have been planning this project using the TPC methodology, and so have been able to optimize our project's schedule and, ultimately, EMV without cutting scope. However, this is not always the case, nor will it necessarily remain the case with our MegaMan project.

In order to justify its inclusion in a project, an activity must add more value than it costs. This might seem like a banal concept, except that we have learned that the cost of an activity is often much greater than the price of the resources it uses; if it winds upon the critical path, its true cost can be greatly increased due to drag cost. The fact that so many projects are scheduled without using CPM almost guarantees that activities with *net value-added (NVA)* that is negative will be included, and even projects that use CPM are prone to such inclusions if they do not incorporate the concepts of:

1. The TPC Business Case
2. Drag
3. Drag cost
4. The VBS and activity value-added

As we said, right now the MegaMan project does not need to prune scope because all of its activities are adding more value than they are costing. This is primarily due to the fact that there is now only one activity with drag—*Design MegaMan* has 1 week of drag at a drag cost (given the

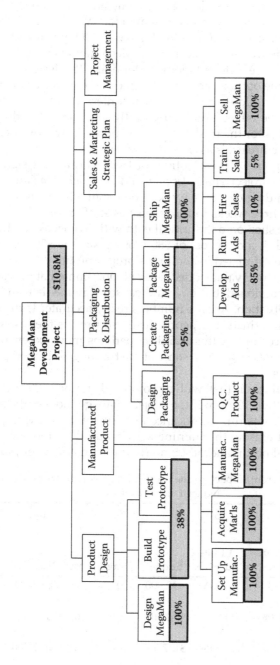

Figure 8.6 Current VBS of the MegaMan project with percentages of project EMV of $10.8 million.

current projected completion date) of $400,000. Add its budget of $187,333, and its net cost is nowhere close to its value-added of (now) $10.8 million. In fact, since *Design MegaMan* is a mandatory activity, we can never cut it from the project; its value-added is always 100 percent of the project's EMV because we can't make and sell MegaMan without designing it (although we may use different methods for doing the design).

What about the optional activities? Three of them are so valuable that it is unlikely we would ever remove them unless we canceled the project. However, three others are of sufficiently low value that they might become candidates for the trashcan at some point. Currently, their status and impact on the project is as shown in Table 8.8.

Currently, all three of these optional activities are adding significantly to the value of the project. *Hire Sales Staff* and *Train Sales Staff* are currently "insulated" from drag cost by a week of float. But what would happen if it took an extra three weeks to hire the sales staff? Suddenly, *Train Sales Staff* would be pushed onto the critical path with two weeks of drag. That would carry with it a drag cost of $800,000 versus the $360,000 that the training has been estimated to add to the project profit.

The negative net value-added should not immediately cause the training to be abandoned, but it should cause an immediate and careful analysis of the situation. Are the numbers really accurate? Is the value of trained versus untrained salespeople really so tiny? Is there some other way we could accomplish this training, perhaps more expensively, but utilizing less time? Can we trim some of the training and accomplish it in one or two weeks?

The key to all this is that, with a project of 1,600, or 16,000 detail-level activities rather than 16, even if the project manager notices the slippage, she will not notice its full significance in terms of project value, nor the likely remedy of erasing the training activity, until it's too late to do so. The training will occur, and other, more drastic, methods of pulling the

Table 8.8 Current Impact on the MegaMan Project of Three Optional Noncritical Activities

1	2	3	4	5	6	7 (4 ± 6)	8 (3 – 7)
Activity	Value-added %	Value-added $	Budget	Drag	Drag cost	True cost	Net value-added
Prototyping (combined)	38%	$4.09M	$42,000	0	$0	$42,000	$4.05M
Hire Sales Staff	10%	$1.08M	$82,500	0	$0	$82,500	$1.00M
Train Sales Staff	5%	$0.54M	$177,500	0	$0	$177,500	$0.36M

project back on schedule will be analyzed. Scope trimming will occur, though often invisibly and in ways that harm both the product EMV and the good reputation of the company. Other projects will be delayed in order to shovel resources over to MegaMan. Finally, the project will probably never regain that lost week.

What is needed is a project management software package that will take the input from the TPC Business Case and VBS and then apply it to the working schedule at the activity level. In this way, if low net value-added (NVA) activities start to slip, the software package would immediately generate an Exception Report to the effect that the following activities have negative (or very low positive) NVAs. Then it would be up to the project manager to analyze the data and implement the most satisfactory solutions.

Unfortunately, no such software exists currently. One would almost think that there is no relationship between project activities and profits.

Summary

So what ABRA has done for us, so far, is to generate a budget for each activity and for the whole project. This is done by pulling the required resources from the resource database and assigning them to the project, activity by activity. The resource database contains cost rates for the usage of each resource, and this allows us to develop the budget for each activity, and for the overall project.

What we now have is a project plan that includes the following information:

1. Total project duration = 28 weeks
2. Total project budget = $4,036,783
3. Project EMV = $10,800,000
4. Project profit = $6,763,217
5. Project DIPP = 2.68

We also have:

- A budget for each activity
- A schedule for each activity
- The precise resources that will be used on each activity

The combination of the budget and the schedule will allow us to generate a tentative *cost accrual schedule,* reflecting how the resource utilization and dollar expenditure would accumulate, *if* the current schedule were implemented. However, we don't yet know if we will be able to implement the current schedule because we still don't know if we will have the necessary resources when we will need them.

We do have the data we need, though, in order to find out. A combination of this schedule for each activity and the list of precise resources that each activity will use provides us with a schedule of our resource needs. Now all we have to do is compare a calendar schedule of our resource needs against the calendar of resource availability—the database, called a *resource library*, that each department should be maintaining. This brings us to the final chapter of the TPC planning process.

chapter nine

Resource scheduling and leveling

In the last chapter, we saw that there are three items of information regarding resource requirements that need to be determined for each activity:

1. **What** resources are needed?
2. **How much** of each resource is needed in order to complete it *within its estimated duration?*
3. **When** will each resource be needed?

The first two of these allow us to develop our budgeting and costing information, but the third is at least as important as the other two. Without it, you can never know that you, in fact, will have the resources you need in order to meet your schedule. And if you don't and if it's *really important* that you meet that schedule, then either the project should be cancelled or the **"What?"** and the **"How much?"** have to change.

The three data elements listed above are entered into a project management software package and matched against the database consisting of the organizational breakdown structure (OBS) of internally available resources, as shown in Figure 9.1.

The database of the internal resources should contain information regarding:

1. **What** resources are available?
2. **How much** of each resource is available?
3. **When** will each resource be *unavailable?*
4. **What is the unit cost** of each resource?

The last item, the unit cost, allows the resource usage to be translated into a budget that is stated in a common (monetary) unit. This not only allows comparison of one type of resource with another, but also the results of one decision with another, in terms that are usually (though not always) directly related to the organization's reason for existence: profit. The fact that companies are often so fastidious about budgetary caps and other cost controlling measures, yet so negligent about project value forecasts, is, on the face of it, a strange dichotomy. Surely *someone* realizes that the two sides of the profit equation are closely linked and *should be* closely

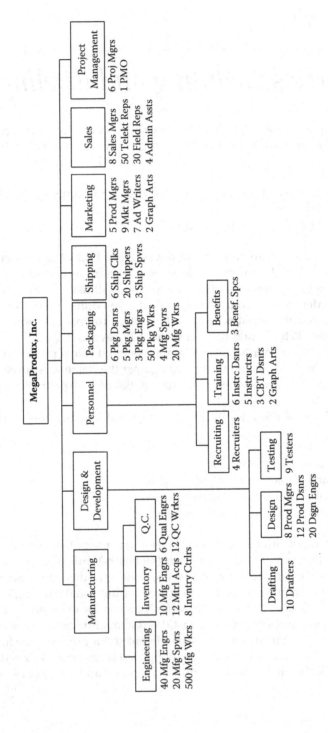

Figure 9.1 OBS of MegaProdux, Inc. showing internal resources (from Chapter 8, Figure 8.1).

linked and should be monitored, from the activity level up, with equal fervor?

The other three items combine to assist us in performing the last of our scheduling functions, which is making sure that we have a work schedule in which the resources will always be there when we need them.

Most modern project management software packages have the functionality to incorporate one or more databases for resource availability as well as multiple project schedules, complete with resource assignment lists. As each project is assimilated, it reserves the resources it requires during the time that it's scheduled to need them. Then, when a new project is entered into the software, the resources that the previous project took are displayed as already assigned and, thus, no longer available. The software then matches the project data against the availability data and does two things:

1. It draws attention, through exception messages or online reports, to resource bottlenecks where there is insufficient resources to meet an activity's requirements during its scheduled time period.
2. It applies a set of alternative methods, some preprogrammed and some user-defined, designed to "get around" the bottleneck. Optional methods might include:
   a. Splitting an activity into one or more parts.
   b. Delaying an activity until sufficient resources become available.
   c. Utilizing a "Priority" field, at either the activity or the project level, to "steal" resources from another activity or project to which they had previously been assigned.

This process of sliding the schedule around in order to reduce a project's peak resource requirements and eliminate resource bottlenecks is called *resource leveling*. The ability to perform this function easily, transparently, and effectively (in the sense of rescheduling activities with minimum delay or extra cost) is a very important feature of a good project management system, and one in which there is substantial variation between packages.

It also is important to be able to level resources across the organization's entire portfolio, on a multiproject basis. Just because another project was entered into the software first and snatched up all the available resources, this should not mean that those resources can't be reassigned to a newer, higher priority project if they are "vital" (and just what *does* that word mean?) to that project.

First, let's look at how resource leveling works (or is intended to work) and then we can discuss how it *should* work.

Resource leveling of the critical path

Two activities, *Set Up Manufacturing* and *Design Packaging*, require the
resources as shown in Table 9.1.

The first hint that there may be a problem comes from the fact that the
two activities are scheduled to occur at the same time and both require
manufacturing engineers. This can be displayed nicely in the Gantt chart
format (Figure 9.2).

Table 9.1 Chart of Showing the Resource Needs of Two Activities in the
MegaMan Project

Activity	Duration	Assigned resources	Work days	Cost/Wd	Cost
Set Up Manufacturing	3W	2 Mfg. Engrs.	30WD	$400/WD	$12,000
		1 Mfg. Spvr.	6WD	$300/WD	$1,800
		4 Mfg. Workers.	60WD	$200/WD	$12,000
Design Packaging	3W	1 Pkgng. Dsnr.	15WD	$400/WD	$6,000
		1 Pkgng. Engr.	15WD	$350/WD	$5,250
		1 Mfg. Engr.	5WD	$400/WD	$2,000

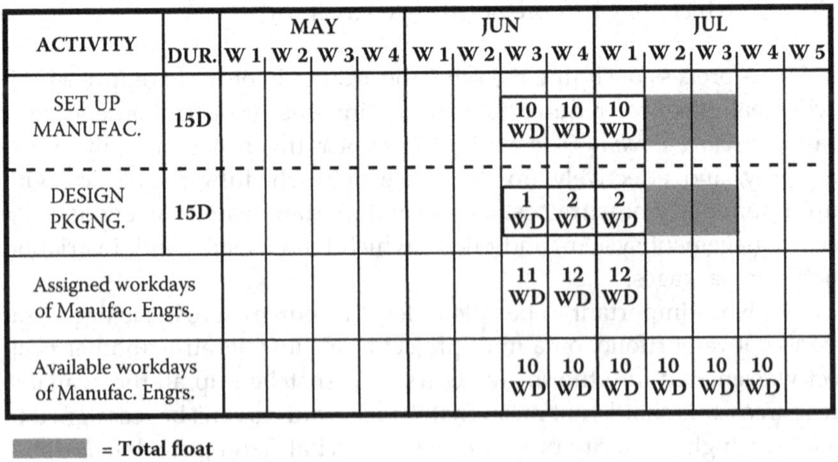

ACTIVITY	DUR.	MAY				JUN				JUL				
		W 1	W 2	W 3	W 4	W 1	W 2	W 3	W 4	W 1	W 2	W 3	W 4	W 5
SET UP MANUFAC.	15D							10 WD	10 WD	10 WD				
DESIGN PKGNG.	15D							1 WD	2 WD	2 WD				
Assigned workdays of Manufac. Engrs.								11 WD	12 WD	12 WD				
Available workdays of Manufac. Engrs.								10 WD	10 WD	10 WD	10 WD	10 WD	10 WD	

▨ = **Total float**

Figure 9.2 Gantt chart of two activities showing required levels of the manufacturing engineer resource versus available levels.

With 40 manufacturing engineers assigned to the Engineering Department, the project manager (and the activity manager) might well feel that there should be no problem getting the necessary resources to perform both activities simultaneously. Unfortunately, this is a dangerous assumption to make. There is a reason the Engineering Department has 40 manufacturing engineers; it's because they are busy. This is exactly the kind of resource and department that desperately needs a carefully maintained resource library, and it's exactly the kind of department that never seems to have one.

Fortunately, we at MegaProdux, Inc. understand the value of resource libraries. Senior management has mandated that every department must have one. So, when we load our project into the software, we receive an exception message, informing us that there is insufficient availability of the manufacturing engineer resource to meet our needs for the two activities *Set Up Manufacturing* and *Design Packaging*.

Many software packages also will allow us to examine the overloaded resources and the bottlenecked activities, first on a Gantt chart as shown in Figure 9.2 and then on a histogram as seen in Figure 9.3.

Because we used the critical path method in scheduling this project, we can readily see that we have good alternatives for solving the problem. The two weeks of total float that each activity enjoys is the key; *how* we choose to employ that float in scheduling the activities and resources is a matter of preference. We can give priority to *Set Up Manufacturing* as shown in Figure 9.4, which allows *Set Up Manufacturing* to finish in

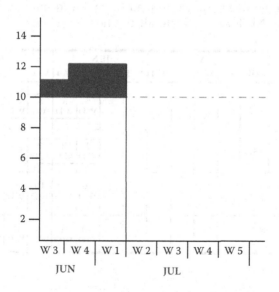

Figure 9.3 Histogram showing required levels of the manufacturing engineer resource versus available levels.

ACTIVITY	DUR.	MAY W 1	W 2	W 3	W 4	JUN W 1	W 2	W 3	W 4	JUL W 1	W 2	W 3	W 4	W 5
SET UP MANUFAC.	16D							10 WD	10 WD	9 WD	1	(float)	(float)	
DESIGN PKGNG.	15D									1 WD	2 WD	2 WD		
Assigned workdays of Manufac. Engrs.								10 WD	10 WD	10 WD	3 WD	2 WD		
Available workdays of Manufac. Engrs.								10 WD	10 WD	10 WD	10 WD	10 WD	10 WD	

▓ = Total float

Figure 9.4 Gantt chart showing resources leveled with *Set Up Manufacturing* receiving priority.

16 days, but pushes *Design Packaging* out to the end of its float (so that it becomes critical), or we could give priority to *Design Packaging* as shown in Figure 9.5, which would leave *Design Packaging* unchanged, but stretch *Set Up*.

Or, we could split either activity in two, or … . Whichever schedule is deemed preferable (and, without additional information, my gut feeling would be NOT to eliminate all the float on *Design Packaging*), the

ACTIVITY	DUR.	MAY W 1	W 2	W 3	W 4	JUN W 1	W 2	W 3	W 4	JUL W 1	W 2	W 3	W 4	W 5
SET UP MANUFAC.	18D							9 WD	8 WD	8 WD	5 WD	(float)		
DESIGN PKGNG.	15D							1 WD	2 WD	2 WD	(float)			
Assigned workdays of Manufac. Engrs.								10 WD	10 WD	10 WD	5 WD			
Available workdays of Manufac. Engrs.								10 WD	10 WD	10 WD	10 WD	10 WD	10 WD	

▓ = Total float

Figure 9.5 Gantt chart showing resources leveled with *Design Packaging* receiving priority.

important factor is that, because of the float, the planned completion date for the entire project will not be delayed and, therefore, its expected monetary value (EMV) will be unaffected. At least thus far. It is possible that there could be a ripple effect, with a successor activity also now being pushed through its float to a period of time when there is another resource bottleneck. That new bottleneck might not be resolvable off the critical path.

However, in general, if two activities are competing for the same resource, and one is on the critical path and one is not, whichever is on the critical path should have priority, delaying the other activity within its float.

Resource leveling on the critical path

What happens when we cannot eliminate resource bottlenecks off the critical path? When, for example, two critical path activities are competing for the same resource, or when the float of the one off the critical path is so small that the resource delay will make it critical?

Let us look at the two prototyping activities, which are somewhat in parallel and share the need for certain resources (Table 9.2).

Again the software generates an exception message: We have a bottleneck due to insufficient design engineers (Figure 9.6).

Build Prototype is on the critical path, and *Test Prototype* has only one week of float. In Figure 9.7, we show how trying to schedule these activities within the current resource limits pushes *Build Prototype* to 30 days beyond its float. The need to fit in 50 days of design engineer work will push out the project completion beyond the target date.

Again, this is two activities, but what would be the ripple effect, over a 1,600 activity project? Only a computer could wrestle through all the possible permutations and give us an accurate picture of what our schedule might be within our resource limits.

Table 9.2 Chart Showing Overlapping Need for Design Engineers on Two Activities

Activity	Duration	Planned weeks of work	Assigned resources	Work days
Build Prototype	5W	Weeks 4–8	2 Design Engineers	50WD
Test Prototype	2W	Weeks 5–6	2 Design Engineers	20WD

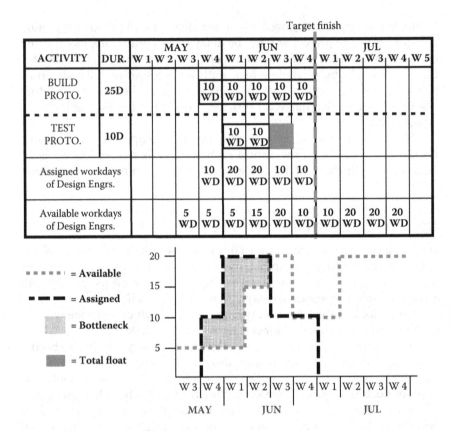

Figure 9.6 Gantt chart and histogram of two activities, one critical, showing required levels of the design engineer resource versus available levels.

Time-limited versus resource-limited resource leveling

Most project management software packages have two completely different algorithms for resource leveling. They are called *time-limited resource leveling* and *resource-limited resource leveling*. Most of the time, there is a "toggle" within the software that allows you to select the algorithm you want to use.

Many software packages can be tremendously helpful in trying to compute the shortest possible schedule. However, a word to the wise: all packages have limitations; none, for example, will pull an activity *earlier* than its critical path method (CPM) early dates. But a savvy project manager might see that a CPM constraint is costing hugely valuable time, rethink how to do the work, and thus avoid a large and expensive delay. Sometimes a slight change in the user-defined parameters, such as

Target finish

ACTIVITY	DUR.	MAY				JUN				JUL				
		W 1	W 2	W 3	W 4	W 1	W 2	W 3	W 4	W 1	W 2	W 3	W 4	W 5
BUILD PROTO.	25D				5 WD	5 WD	10 WD	10 WD	10 WD	10 WD				
TEST PROTO.	10D						5 WD	10 WD	5 WD					
Assigned workdays of Design Engrs.					5 WD	5 WD	15 WD	10 WD	10 WD					
Available workdays of Design Engrs.				5 WD	5 WD	5 WD	15 WD	20 WD	10 WD	10 WD	20 WD	20 WD	20 WD	

▓ = **Total float**

Figure 9.7 Gantt chart of two activities, one critical, showing schedule slippage beyond target date.

priority assignment, can cause the computer to generate a schedule that is much more or less efficient. That is because the number of variables on anything more complex than a very short project can introduce so many possible permutations that it would take hours for even the fastest computer to optimize fully.

In talking about time-limited versus resource-limited leveling, it is important for the software user to realize that these are not really two different tools, nor two different schedules. Rather, they are two alternative *pictures* of the same data set, treated first in one manner and then in the other.

In the first instance, we will toggle the software so that it produces a *time-limited* schedule. What this means is that the computer may delay activities within their float, split activities in two, and obey any of the preset parameters as defined by the user. The computer may even schedule activities for which there are not enough resources, but the one thing the computer *may not* do is delay the project beyond the completion date that the user has specified.

Now, although the software manuals will not state this, and the software itself will allow you to specify any date for completion, *it is crucial that the initial snapshot of the time-limited project schedule be based on the project completion date from the CPM schedule.*

Why? Because this is the date for which we know we can plan if we don't run into resource shortages. Therefore we need to know what the completion date will be if we don't have to delay the project because of shortages versus if we do, because resource shortages are things we might be able to do something about.

So now we have a schedule generated by the computer dealing simultaneously with all the resource restrictions, on all the different activities, but without pushing any activity beyond its late finish date.

Next, we toggle the software so that it will produce a *resource-limited* resource schedule. Just as it sounds, this will be a schedule in which the software will once again try to resolve resource bottlenecks, but this time it will not schedule any activity unless and until the resource requirements are fulfilled. That can mean only one thing: If the resources for an activity do not become available within that activity's float, *then it has to be delayed beyond its float,* and that means that the project completion date may have to be delayed.

You get your pick: A or B. You can go out and get the additional resources you need and finish the project on time, or you can hold the line on the budget and finish it late.

I have found that the best way to look at this information is in the form of two histograms, one placed immediately above the other so that the schedule implications can be easily seen. Figure 9.8 displays these types of histograms.

In one case, we need more resources in order to meet the project completion date. In the other, the resource limitations are accepted as rigid, and the project completion date slips. Again, this is A and B. If we like, we can analyze A1 and B1, where we add a little more of this resource during one period, or A2 and B2, with a little more of that resource during another period. But when all is said and done, the resource-limited resource schedule will be our project's working schedule. Whatever resource limitations remain in our project plan at the end of this process, we are going to have to live with them, and the project schedule must reflect those limitations, because they are *reality,* and the project manager must always make of reality an ally, not a foe. That means *using* reality to *improve* reality.

All the information and formats above have long been a standard part of the traditional project management methodology (even if often ignored in practice). For decades, project managers have been trying to acquire additional resources by showing senior management the kind of data displayed above. "If we don't get two more electrical engineers (or one more plumber, or programmer, or widgeter), the project's going to slip by three weeks." Most of the time the response is something like: "Your budget's already too much. You're going to have to make do without. Do the best you can."

Is it that they don't believe the data? Perhaps. However, if the project has been planned as we have shown, utilizing work breakdown structure (WBS), CPM, activity-based resource assignment (ABRA), and all the other standard project management techniques, then the data, while not infallible, are certainly reliable. What is missing, however, is crucial

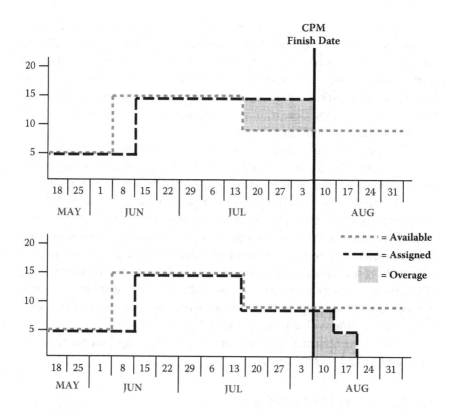

Figure 9.8 Histograms of time-limited and resource-limited resource-leveled schedules.

information that will tie together the two sides of the issue in a way that will allow comparison and decision.

On the one hand, the project needs more resources if it is to finish on time. Those resources have an important negative impact on the project (and on the entire organization)—cost. Measured in dollars.

On the flip side, if the project does not get the resources, it will not finish on time. Time is measured in days and weeks and months, and time is merely an abstraction. Corporations don't work for time, they work for dollars. Any time that a corporate disagreement occurs, with one side arguing in dollars and the other in time units, the side arguing dollars is going to win.

This is, and always has been, the weakness of traditional project management. It has failed to provide true cost/schedule integration, and it has failed to justify itself, because it deals in time and not in dollars. The value/cost of time, left unmonetized, becomes what is called in economics an *externality*. Sure, at some level every corporate executive appreciates that time is money, but *how much* money? And, how deep in their gut do they appreciate it? The cost of one electrical engineer for three weeks is $5,000. You can see that money, touch it, count it. There it is, in black and white on the spreadsheet, with a $ in front of it. What is a three-week delay? You can't see it, and you can't touch it, and the fact that everyone knows it may be worth a helluva lot more than $5,000 doesn't mean a thing if it ain't got that $ and number right there on the spreadsheet.

That's the big advantage of Total Project Control (TPC). The TPC Business Case allows the schedule overrun to be monetized, based on the resultant reduction in expected monetary value. In previous chapters, we saw how the delay cost of the project completion could be charged down to the activity level through drag and drag cost. Now, with the resource-limited schedule, we have further delay costs. These delay costs should be charged down both to the activities being delayed and, even more important, to the resource shortages that are causing the delays. In other words, TPC will allow us to list the cost in lost dollars resulting from a resource shortage on the same spreadsheet as the electrical engineer's $5,000, also in black and white and with a $ before it.

Then we can subtract one number from the other and see what action is preferable, and by how much.

The CLUB (cost of leveling with unresolved bottlenecks)

The difference between the project completion date of the CPM schedule and that of the resource-limited resource schedule may be three weeks. However, if the project is our MegaMan Development Project, and the two dates are Week 28 and Week 31, the difference is also $2.8 million: $400,000 apiece for Weeks 29 and 30, and $2 million for Week 31. This difference is the result of resource leveling without first resolving the resource bottlenecks. So the TPC term for this reduction in project EMV is the cost of leveling with unresolved bottlenecks. It is the CLUB to be used for getting additional resources—pay now or pay later.

The CLUB is based on the reduction in value of the entire project due to resolving all the resource bottlenecks. If we are employing ABRA, though, resources are not assigned to the project, they are assigned to each activity during the time that the activity is scheduled to occur. Therefore, just as we charged the CPM schedule delay cost down to each activity's drag, so too do we need to charge the project's CLUB down to the activities and resources that generate the delay.

Resource schedule drag

On the CPM schedule, drag is a function of the logic of the work, i.e., the way the work must be done. The drag calculation quantifies the amount of time that could *potentially* be saved by eliminating an activity or by reducing its duration to zero. This is of great value when trying to determine *where*, and *by how much*, one can shorten the project duration on any given activity.

When we load resources and generate the resource-limited schedule, however, another delaying factor comes into play; delay due to the lack of sufficient resources on a timely basis. This type of delay also should be tied to the activity that is being delayed. Although when figuring out where and by how much we can shorten the project, we need to take into account both types of delay (in combination), ultimately we must be able to separate out the two different types of delay, because one can be addressed by adding more resources and the other cannot.

What this means is that on the resource-limited schedule, we need to quantify two different types of drag. The first is the combination drag or what we will call *resource schedule drag*. It is the same type of drag as we had on the CPM schedule. It is the amount of time that could *potentially* be saved by shortening an activity before the critical path changes. The actual number (indeed, the entire critical path) may change in going from the CPM to the resource-limited schedule, but it is still calculated in the same way, i.e., by determining the total float of the parallel activity with the least total float.

Once the raw total of each activity's resource schedule drag has been computed, we then need to determine how much delay is due to which of three different causes:

1. Delay due to the logic of the work, i.e., *CPM schedule drag.*
2. Delay due to other, ancestor, activities that unavoidably push out the schedule of the successor.
3. Delay due to *this specific* activity having to wait for resources, which we will call *resource availability drag* or *RAD;* if the activity is removed, the resource delay will go with it.

Resource availability drag (RAD)

It is this third type of delay, the resource availability drag, that we particularly need to isolate, because it is the one we can really do something about. Furthermore, by tying an activity's RAD to the CLUB, we can see how much it would be worth to reduce or eliminate the portion of the drag due to resource availability. That also would allow us to arrive at the maximum *availability premium* that it would make sense to pay for a given resource.

The formula for calculating the RAD is complex and not one than can easily be used mentally. But for a computer, it would be simple and very valuable. Below, I state the formula for computing the RAD. If you are not interested in how the formula works, but merely in the value of its output, feel free to skip it. Here are the three really important things to remember about the RAD:

1. An activity can only have RAD if it has resource schedule drag. The fact that an activity is delayed because of resource availability does not give it RAD *unless* the delay pushes out the critical path.
2. An activity can never have *more* RAD than the amount of its resource schedule drag. RAD is a subset of resource schedule drag. If Activity X is being delayed by five weeks because its resources aren't available, but the parallel path has only one week of total float, then Activity X has only one week of resource schedule drag and, therefore, *cannot* have more than one week of RAD.
3. As always, drag is found *only* on the critical path; an activity may not have resource schedule drag even though:
   a. it was on the critical path, with drag, on the CPM schedule, and
   b. it is further delayed on the resource schedule.
   If another path on the resource schedule is being delayed even more (so much more as to negate the total float that it had on the CPM schedule), then the old CPM critical path activities no longer have drag, they are no longer delaying the project, so you would gain no time by shortening or deleting them.

The formula for computing RAD

The RAD for Activity X can be calculated by using the following formula:

Resource Availability Drag =	Late Finish (resource-limited schedule) minus Late Finish (CPM schedule) minus (largest difference between any predecessor's CPM Late Finish and resource-limited schedule Late Finish) or the resource schedule drag, whichever is less.

The first part of the formula measures the amount that the activity is being delayed in going from CPM to the resource-limited schedule. The second part of the formula determines the delay due to lack of resources on this activity by subtracting out the amount of the above delay that was caused by ancestor activities. The third part of the formula ensures that the amount of time that will be gained on the project completion date is not overstated. The resource delay may be reduced from 10 days to 0, but if this only shortens the project by 2 days, 80 percent of the reduction would be of little value.

The value of RAD

Resource availability drag is a vital metric to the project, the functional department with insufficient resources and the overall organization, because it is a delay factor that we can do something about. Much of the time, we might be able to correct the problem right on this project and activity. If the RAD due to insufficient programmers is two weeks, and the project's EMV is being decreased by $50,000 for each week of duration, that computes to $100,000 that we would save by removing the RAD. That makes $99,999 we can spend to get additional programming help and still be one dollar better off than we were before. Anyone know of a programmer who can be hired for $49,999.50 per week?

I bet you do. In fact, I bet you know a programmer who would work for two weeks at a mere $10,000 per week, or $2,000 per week. $50,000 is simply the *maximum* availability premium (MAP) we should pay. When quantified in the right way, on a project delay and EMV basis, the resource availability premiums are almost always so huge that it makes the whole process seem ludicrous. (After all, we aren't *really* going to pay a programmer $50 grand a week. That would blow up our salary structure and, ultimately, our entire organization. Everyone would soon be blackmailing us for exorbitant wages.) However, those kinds of numbers are real. This is what it's really costing a project like this not to have sufficient programmers available.

And that's just one project. Where else is the lack of programmers hurting us? How many projects across the entire organization would finish one week or two weeks or three weeks earlier if we had one or two or three more programmers? How much do those weeks add up to, in dollars, in the course of a quarter? In a year?

This is where the functional manager comes into play. Functional managers are responsible for maintaining the resources that support the projects, and functional managers usually find themselves with departments and subordinates that are grossly overworked. Every time a functional manager tries to hire additional resources, he is told that "we've got to hold the line on head count." So the head count numbers keep becoming even more grotesquely unbalanced.

One trouble is that most functional managers really don't understand project management. WBS, critical path, total float; that's all project manager stuff. When corporations run training seminars on project management, the functional managers don't even attend. After all, they are not project managers. They don't have to manage projects. No, they just get to destroy projects because they have never been educated as to the relationship between the project and the resources that they control. They don't know how to get the project data they need in order to justify the additional resources that are required to better support the projects.

One company with which I have worked is an international organization that puts communications satellites into space. Every penny of revenue that this organization generates is related to projects: building satellites, building ground stations, developing telemetric software, changing orbits, selling broadcast time, etc., etc. Yet, until very recently, this company did not use CPM; did not assign resources to projects, never mind activities; and never charged work hours to projects or activities. The board of directors had mandated a budget cap, and additional expenditures and new hires had to be rigorously justified. As a result, everyone throughout the organization complained bitterly about being underresourced.

What is wrong with this picture? How can you possibly justify additional resources in a project-driven company when you have absolutely no data to indicate what impact the lack of such resources is having on projects? Never mind the fact that the data was not quantified in dollars; it wasn't even quantified in time units. Two departments, Human Resources and the Controller's Office, understood the problem and attempted, despite great resistance, to introduce project management techniques and project-based time reporting. Which individuals were the most resistant? The functional managers, of course. Because they could only see that more data would make the inefficiencies of their departments more visible, not how these project management techniques could benefit them.

The resource-leveled schedule for the MegaMan project

The resource bottlenecks due to the lack of sufficient design engineers for the prototyping activities causes the *Build Prototype* activity to take six weeks instead of our estimated duration of five weeks. Worse, when we look at the entire resource-limited schedule, it turns out that we have slipped not one, but five weeks. We knew when we went to the DRED (doubled resource estimated duration) for the *Package MegaMan* activity that we were going to need additional resources. Well, they have not yet been hired. Additionally, the resources to *Ship MegaMan* in one week are there for Weeks 28 or 29, but thereafter they will be needed on other activities, forcing *Ship MegaMan* back to its original two-week duration. Our resource-limited schedule looks as shown in Figure 9.9.

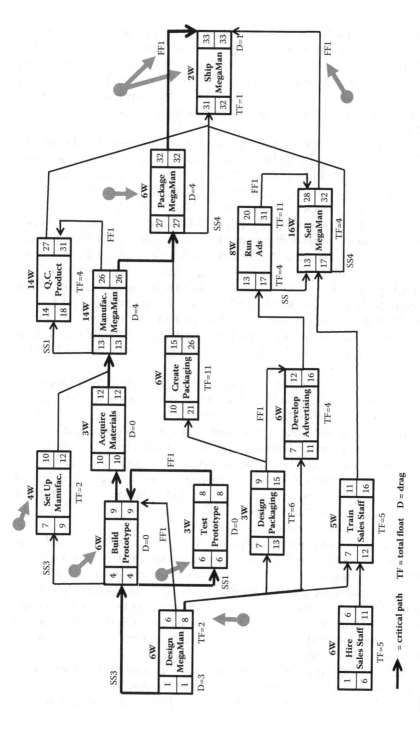

Figure 9.9 Resource-limited resource schedule for the MegaMan Development Project.

Notice that the duration of *Package MegaMan* is back to six weeks, its start-to-start (SS) lag with *Ship MegaMan* is four weeks, and the two free float (FF) predecessors for *Ship MegaMan* again need a week of lag. Additionally, *Test Prototype* cannot start until Week 6, a RAD of one week, and so its float has disappeared and it is now critical. As we discussed earlier, *Set Up Manufacturing* has had its duration increased by a week. *Design MegaMan's* start remains critical, but its finish now has two weeks of total float.

If we have to adopt this resource-limited schedule, our duration will be 33 weeks and our project EMV will decrease to $4 million, a point at which our DIPP (Devaux's Index of Project Performance) would indicate that it is no longer worth undertaking. On the other hand, it would be worth $6 million to shorten it by three weeks, and another $800,000 to get back to where we *know* we can be without resource delays, the CPM schedule duration of 28 weeks. Surely we can use some of this money to find a way to shorten the schedule.

It's time for a meeting. All the data we have assembled thus far show that the net result of our company being short one design engineer and 10 packaging workers, for the few weeks that we need them, will be a profit reduction in excess of $6 million. The activity manager for *Package MegaMan* shrugs, and reminds us that we had said that we would get the approval for hiring the additional packagers through the Human Resources Department. "I've told them we need more people till I'm blue in the face. They won't budge."

A quick phone call gets the director of HR to drop in on the meeting. Once again the data are laid out. The lack of 10 packagers for three weeks is about to cost us $6 million, we explain.

The HR director looks at our charts. Clearly, he is concerned that the blame for lost revenues is going to be laid at his doorstep. He would like to hire more packagers, he explains; several other project managers have been asking for them as well. However, there just aren't any packagers available out there. The economy is really good right now, and it's difficult to attract even unskilled workers for $10 an hour.

Fine, we reply, let's offer $20 an hour.

We can't do that, comes the response. When you add in benefits and other overhead, it would push the cost of a packager from $150 per day to almost $300. It would throw the entire corporate salary structure out of whack.

At this point, we go to the white board and draw a vertical line down the middle. On one side we write the heading: "Cost Due to HR's Refusal to Hire 10 Additional Packagers for Three Weeks." On the other side, we write: "Cost of Paying 10 Additional Packagers for One Year." Then, on the first side we write "$6,000,000."

Now, we offer the HR director the marker and ask him to write in the cost of the packagers. He doesn't do so, but it's okay; the point has been made. When the meeting ends half an hour later, he has agreed to work with the packaging manager to ensure that sufficient temporary packagers are hired to complete the activity in three weeks. He also has agreed to approve the hiring of an additional design engineer, which will allow the prototyping activities to return to their CPM durations.

We, in turn, have agreed to charge the project budget for 12 weeks' worth of 150 percent of the burdened wages of each of the 10 packaging workers, plus the design engineer. These will be charged to the specific activities under the line item "Availability Premium." The 150 percent is to make sure the resources can be obtained. The 12-week charge is in case the functional departments are not able to keep the additional employees sufficiently busy and have to lay them off with appropriate separation packages. The charges work out to $27,000 for the packagers and $6,300 for the design engineer.

With the *Ship MegaMan* activity once again scheduled for Week 28, the resources are again available to complete it in one week. We are back to our 28-week schedule, and to our $10.8 million EMV. Not bad, for additional cost of $33,300.

Before we go home that afternoon, we send an e-mail to the CEO explaining how "the proactive support of HR is saving the organization $6.8 million." And, of course, we cc the HR director.

I am only too well aware that the process does not always go that smoothly. There are even occasions when $6.8 million opportunities must be passed up, due to cash flow or other considerations, but, under any circumstances, the numbers that are involved must be computed and made crystal clear. To make a decision to forego $6 million is one thing, to do so without even recognizing the issue is gross incompetence. Unfortunately, it is precisely the sort of incompetence that the corporate world engages in thousands of times a year.

Rightsizing a project-driven organization

- Project delay costs money by reducing the project EMV.
- Project delay can be assigned to specific activities, through CPM and drag.
- The difference between an activity's drag on the CPM schedule and on the resource-limited schedule is RAD, the increased drag of an activity due to resource unavailability.
- Therefore, through ABRA, both project delay *and its cost* can be attached to the resource whose lack of availability is its cause.

This can be done on one project, and on all the projects in the organization that use that resource. However, that information needs to be collected at the resource level, which is to say by the functional department. By maintaining the resource library and assisting project managers in resource scheduling and leveling, functional managers can collect data on the schedule impact of the scarcity of each resource. Through the projects' TPC Business Cases, this impact can be monetized, and additional resources and availability premiums justified. Ultimately a staffing level can be attained that is optimized, on the basis of profits, for each resource.

It is crucial to realize that these are entirely different definitions of optimized and rightsizing. They are definitions driven by the total, integrated nature of a project, and of a portfolio of projects. They are impossible to codify without an integrated concept of the project, a concept that has at its root the monetized value of the scope.

Previous approaches to rightsizing included some idea about keeping people busy. If people weren't busy all the time, resources weren't being used to full capacity and, thus, were being wasted. This concept is an anachronism, left over from the days when business consisted almost exclusively of steady, repetitive work. With projects, that paradigm is out the window. Projects don't have steady and level work requirements and, therefore, don't have steady and level resource requirements. Instead, they have a dominating need to finish as soon as possible and a subsidiary need to have those individual items of work of which they are comprised occur at specific times. That means that they need to have the resources to do those individual work items at those times. If those resources aren't there, that original dominating need is compromised.

This may seem relatively obvious, but now there is a corollary, one that many managers, and all those management consulting companies that specialize in downsizing, refuse to recognize: As long as you have got those resources when you need them, it doesn't much matter what they do when you don't.

Henry comes to work every morning at 8:30, puts his feet on his desk, slips on the headphones from his smartphone, and pulls out a good book. About every two weeks, a project lands on his terminal. He grabs his mouse and keyboard, works feverishly until he is finished, forwards it to the next person, then picks up his book again. Are there any Henrys where you work? *I* certainly don't ever see them. In most companies, they would be fired the first week. Certainly, they would be laid off as soon as the first management consultant walks through. *Yet there is nothing whatever wrong with Henry's behavior.* Maybe the good book is about HTML programming, or project management, or that new corporate concept, the learning organization.

Under any circumstances, Henry's value must not be determined by what he is doing when he is *not* working; his value comes from what he is doing when he *is* working. If we decide that we don't need Henry goofing off four days out of every five and therefore, give him the pink slip, we also lose him for that fifth day when he works. And to *not* have him there when our projects need him might cost us 20 days of resource availability drag over the course of the year, and those 20 days might be worth $1,000,000. But that's okay, right? Because at least we have got Henry's salary off our books, which, including overhead, is $200,000. Right? No?

Of course, if we can get Henry's $1,000,000 of work for less than $200,000, that's fine. Perhaps we can lay him off and then utilize him as a consultant one day a week. We will undoubtedly have to increase his hourly wage, but since we won't have to pay him when he's not working (we will also save something on health insurance, workmen's comp., etc.), we will be much better off.

This is precisely why the labor force of consultants, freelancers, stringers, temps, etc. has burgeoned so greatly in the past four decades. Most owe their careers to projects, the numbers of which also have burgeoned. Projects make temp workers extremely valuable, because they can be brought on board just when they are needed, targeted to the right work, and then their cost eliminated when they are no longer required. And it's worth paying a premium for this flexibility and targeted effort, so the busy freelancer's income can be significantly larger than that of the permanent employee.

Unfortunately, sometimes that freelancer you have been counting on can grab an assignment with someone else's project, just when you really need her. By the time you discover this, and either replace her or make her an offer she can't refuse, your project has slipped by two wee ... er, $150,000.

Some jobs you just *can't* use temps for. The worker needs to be thoroughly familiar with the organization, or the specific product, or the type of product. If he isn't, there is a ramp-up time that must be factored in, and, for some types of work, there simply is not sufficient supply of that resource in the labor force to be able to depend on it being available on a temporary basis when you need it.

In such cases, you have to maintain the resource on staff, even if you only utilize it a small percentage of the time. The downside of not having it available, reductions in EMVs all across the board, is just too great. The dollars you shell out to Henry for him to sit and listen to Mozart are just a different form of that availability premium you pay to keep that resource on call. If it grieves you, or Henry's colleagues, or senior management, to watch Henry sitting there every day, tell him to go sit at home instead. Just make sure his cell phone will always be on him so you can get hold of him as soon as you need him. And, whatever you do, don't irk him too much. If you do, he's liable to realize that the four extra days a week he spends listening to Mozart could very profitably be "rented out" to other employers. A consulting business might allow him to retire years earlier and devote all the more time to Mozart.

HR and the CLUBs

Every three or six months, the Human Resources Department of project-driven companies should assemble a CLUB meeting. This would be for all the functional managers, who would bring the data regarding the collected CLUBs of each of their department's resources during the previous period. These should be arranged into Pareto charts, listing in descending order the delay costs due to the shortage of each resource. HR should then present the information to senior management, showing which resource shortages, across the entire organization, are costing the most, and outlining a plan for lessening the problem. This plan should include:

- Which and how many of each resource will be added to the permanent staff?
- What will be done to make sure that temps and freelancers are available when needed?
- What availability premiums will be paid to accomplish each of these goals?
- What will the long-term dollar benefit be as a result of HR's proactive vigor?

There are few things that HR could do in a project-driven organization that would be of greater "visible" value.

In general, HR, like functional managers, does not have sufficient knowledge of how projects and project management work to be able to support such an organization properly.

Sometimes this ignorance can be extremely damaging. A classic example occurred several years ago in a very large international computer company. This company had announced that it would be bringing out a new version of its software by the end of the year. Knowledgeable individuals in the computer industry and on Wall Street smiled knowingly, remembering that each previous release of the software had come out months late. Not this time, said the company, we are going to have it in the stores by December 31. Word was passed to all key members of the project team: Don't plan on seeing much of your families after October; this project is too important to slip.

Unfortunately, the corporation had been suffering through some hard times and had been watching its market share and revenues slip for many months. Cost-cutting mandates had gone out to all parts of the company; travel was being restricted, supply cabinets were being locked, and all the other classic symptoms of a management that doesn't know what to do were evident.

Late that summer, a memo was released from HR that was distributed internationally. It read:

> We are aware that in past years, many employees have elected not to use their vacation time, allowing it to accrue instead. Unfortunately, this is a cost liability on our books, and, in these financially strapped times, we cannot allow this to occur any longer. Henceforth, employees will only be allowed to carry two weeks of vacation time into the new year. All previously accrued vacation time must be used up by the end of this year or else it will be forfeited.

You can imagine the effect of such a memo on the critical path activities of all the company's projects, including the software release. Employees who had not taken vacation in years did so, starting the next Monday, and when they returned in early January, their projects had already slipped by weeks. The new software version came out months late, as usual, just in time to be destroyed by its major competitor. The cost, in both revenues and market share, was considerable; in terms of credibility and stock price, it was even worse.

Sure, the vacation time was a liability on the corporate ledger. But, this is a classic example of snapping at pennies while dropping millions, and is typical in organizations that have zero knowledge of the implications of project work and project management techniques.

The concentration on "work" is an essential part of the project management process. This can sometimes cause those whose job it is to deal with the people side of things, both functional managers and HR, to feel diminished in significance. Nothing should be farther from the truth. If performed properly, project management should actually increase the importance of resource management. Not only are resources the only way to get the work done, but projects introduce a tricky new dimension in managing them successfully: *time.* It's fine and dandy to know that you have 10 good people in your department. But:

- *When* do you have them? And when do you not have them?
- When is each assigned to other projects?
- When will only seven be needed, and what will you do with the other three at that time?
- When, and for how long, will you need 12 people?
- What will the cost be, to projects, to your own department, and to the entire organization, if you don't get them? If you only get 11?
- How much of an availability premium should you be willing to pay?
- How long before the extra two people are actually required will you need to hire them and train them?

All of these questions are TPC issues. They all require that resource managers understand project management. They all require a well-maintained resource library, and they all require project managers who utilize a TPC Business Case and critical path planning, and include the functional departments in the project planning process.

Multiproject resource scheduling

Among project management software packages, the ability to perform multiproject resource leveling is fundamental. However, most software packages are abject failures at performing multiproject resource leveling. The reason is simple—they don't have the information to level projects properly.

Many years ago, when I was first delving into project management theory, I worked for one of the leading distributors of project management software. At the time, they were developing a new product for the Unix environment. I was still quite naive about, and intimidated by, the software, and could not understand why changing the "Priority" field should randomly reduce or increase the resource-leveled duration. I went and asked the product designer for the software if the new leveling algorithm would produce an "optimized" schedule. He looked at me pityingly.

"Steve, what is an 'optimized' schedule anyway? Are you talking about optimizing for time? Is the shortest schedule what you want? Or are you talking about cost? Is the cheapest schedule the optimized schedule? Don't you see? It all depends on what you want."

I walked away from that meeting feeling a bit embarrassed, but, even more, confused. I sensed that somehow there should be more to it than that. A human being can tell which project schedule is better, so a computer should be able to figure it out as well. Yet, I could not decide. Was it shorter schedule or less cost that should take priority?

That problem idled at the back of my brain for several weeks. Then, slowly and with much associated self-doubt, the answer took form: revenues—profit—cost reduction—*value*. *That* should be the criterion for determining the best project schedule. *That* was the whole reason we were doing the bloody project in the first place. But *that* was also an item of information that the software did not have. Project management software deals with cost and schedule. Value just isn't important on a project.

On a single project, you can often get away with the assumption that a shorter schedule means greater value. Sometimes this assumption is invalid, but overall it's not a bad rule of thumb. (The idea that lower cost means better value is another story altogether.)

When the issue is multiproject resource leveling, though, the "short is better" rule becomes incomprehensible. Shorter *which*? Shorter *what*? Shorten one project and you lengthen another. Complete them *all* in less time? But each has a different delay penalty or acceleration premium.

Some may be able to slip a month with only slight impact, while others may cost you half-a-million bucks a *day*. Applying all the different parameters and constraints is a job for a computer, if there ever was one. Yet, project management software doesn't accept that kind of input.

What about the caveat that a computer, *even if the right data were input,* doesn't have the capacity to compute an optimized schedule. If a computer that can beat Magnus Carlsen at chess cannot figure out the best way to distribute the resources, what chance do *you* have of doing it? (Okay, so you have never in your life lost a game to Magnus. Still … .)

The computer may not be able to compute an *optimum* schedule, but it can generate a damn good one. If the right data are entered, and the software has the right algorithm to manipulate the right metrics in the right way, then, given enough time, it should be able to isolate the multiproject schedule that would get the highest score. Or maybe the 10 highest scores. It also should be able to tell you exactly what those scores are.

What units should those scores be in? If you have read all these pages, and still cannot guess the answer, I'm moving back to the West Indies and becoming an obeah (voodoo) man. Dollars, of course; measuring the expected monetary value across all the projects, adjusted for net present value, and minus costs. Profit, quoted either in raw dollars or, because it's simple and convenient, the Simple DIPP.

Every night before he or she goes home, the portfolio manager of all the projects in the organization should go to the computer and access the project management software application with its entire portfolio of current and forthcoming project files. Then a click of the mouse, and the software should spend the night running through the activities and resources, assigning first two programmers to this project and one to that, then two to that and one to this. At 8 the next morning, the portfolio manager should be able to see the current schedule and its EMV numbers, as well as the top three new schedules the software has been able to generate, with their EMV numbers. If there's not much of a muchness, the portfolio manager can decide to stick with the current schedule. Indeed, if there *is* a chasm between the different schedules, the portfolio manager may still eschew the changes. Grandmaster chess players don't just accept the analysis of their assisting computers at face value, they check to see if that analysis makes sense. Management remains the province of human beings. But a computer that has the right data and algorithms can be an indispensable tool.

chapter ten

Tracking and controlling the project

Theory divides the process of managing a project into two parts: planning and tracking. Project management, it is said, is about planning the work and then working the plan.

There is value to this way of visualizing a project. However, in another sense, it introduces a false dichotomy. The planning process does not stop when the project begins. In fact, the tracking process is *more* about planning than anything else. It's about *replanning*. Over and over and over again, whenever a variance from what was previously planned occurs. Those who would pretend that planning a project is to create a rigid documentation that is not supposed to change when the facts change are simply ignorant of the methods of project management.

There is one big advantage to replanning when the original plan was in a traditional project management format, which is we almost never have to go back to square 1. The original plan (remember that stuff about A–I–M F–I–R–E in Chapter 2?) is a vital tool for facilitating the replanning. Indeed, making the process of replanning easier is a major goal of the original planning process, and a big part of the reason for such formats as work breakdown structure (WBS), critical path method (CPM) network, activity-based cost (ABC), etc.

The "A" of A–I–M F–I–R–E, you may recall, stands for *Aware*: awareness that a change has occurred. That means a change from the current plan. The original plan needs to be good, and detailed, and then needs to be closely tracked.

We now have the elements of the plan for the MegaMan Development Project. Next, we need to submit all of these elements to the project review process for the next phase, prior to the final "gate" of project approval. This, you may recall, is the third gate of the process. The first gate approved funding through completion of the project plan. That is where we are now. In order to proceed with the project, we need further funding, and in order to get that funding, the entire plan, including budgetary requirements, must be submitted.

Figure 10.1 shows all the three phases, through the gate signifying approval for funding of the detailed plan.

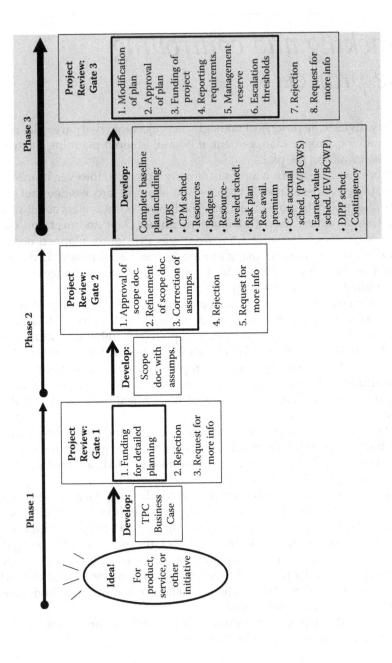

Figure 10.1 The project review process for gating and funding.

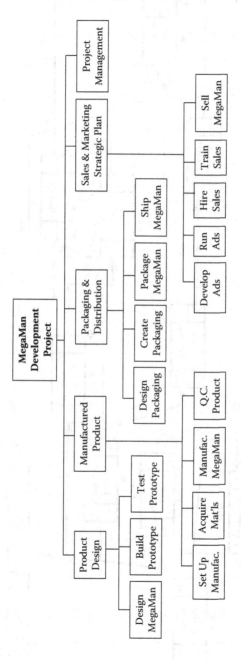

Figure 10.2 WBS for the MegaMan Development Project.

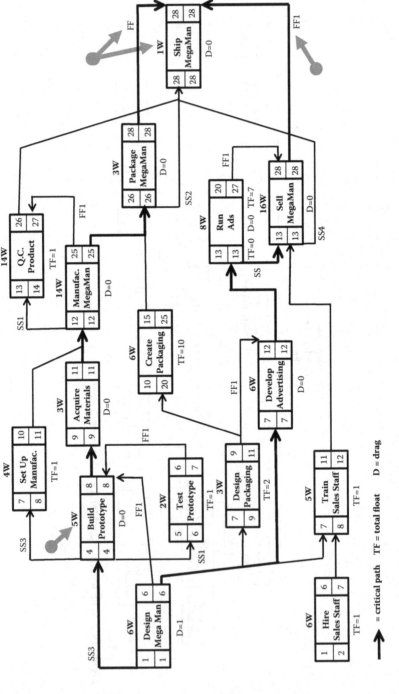

Figure 10.3 Baseline schedule for MegaMan Development Project.

The elements of the plan that must be approved and that will be used to track the project include:

1. **The TPC Business Case**:
   $10 million if completed by the end of Week 30.
   $2 million less for each week later.
   $0.4 million more for each week earlier.
2. **The project work scope document, including the assumptions appendix.**
3. **The project work breakdown structure.**
4. **The resource-limited schedule.**
   Total duration of 28 weeks for expected monetary value (EMV) of $10.8 million. Through the securing of resources by the payment of availability premiums, the project baseline schedule is essentially the same as the CPM schedule. The only resource delay is off the critical path, where the *Set Up Manufacturing* activity has been stretched from three to four weeks, and its total float reduced from two weeks to one.
5. **The activity-based budgets** (Table 10.1).
6. **The project budget** (Table 10.2).
7. **The project's Starting DIPP**: $10,800,000 divided by $4,070,083 = 2.58
8. **The project value breakdown structure** (Figure 10.4).

There are two other elements, both traditional project management documents, that should be a part of the final plan. The data to assemble these two documents have already been generated through the planning process. They include:

9. **The cost accrual schedule.**
   Because we have planned the cost that we expect each activity to incur, and we have scheduled each activity, we also know when the cost for each budget item will be incurred and how that cost will accumulate. We therefore can assemble two histograms:
   a. A periodic cost accrual histogram, showing how much cost will be incurred in each time period.
   b. A cumulative cost histogram, showing how that cost will accumulate as the project work is performed.

Most project management software packages will extract the data from the plan and allow us to generate both these cost functions on the same histogram display, simply using different *x*-axes for the different dollar functions.

An example of such a histogram is shown in Figure 10.5.

Table 10.1 Activity-Based Budgets for the MegaMan Development Project

Summary activity	Activity	Labor budget	Availability premium	Total budget	Summary budget
Product Design			$6,300		$235,633
	Design MegaMan	$172,000	$0	$187,333	
	Build Prototype	$140,500	$6,300	$34,967	
	Test Prototype	$21,500	$0	$13,333	
		$10,000			
Manufactured Product		$397,000			$1,323,333
	Set Up Manufacturing	$25,800	$0	$86,000	
	Acquire Materials	$1,600	$0	$5,333	
	Manufacture MegaMan	$357,000	$0	$1,190,000	
	Quality Control Product	$12,600	$0	$42,000	
Packaging and Distribution		$133,350			$293,700
	Design Packaging	$13,250	$0	$26,500	
	Create Packaging	$51,000	$0	$102,000	
	Package MegaMan	$63,000	$27,000	$153,000	
	Ship MegaMan	$6,100	$0	$12,200	
Sales and Marketing Strategic Plan		$365,250			$2,226,250
	Develop Ads	$20,250	$0	$101,250	
	Runs Ads	$5,000	Plus $400K for Ads	$425,000	
	Hire Sales	$16,500	$0	$82,500	
	Train Sales	$35,500	$0	$177,500	
	Sell MegaMan	$288,000	$0	$1,440,000	
Project Management					$100,000

Table 10.2 Project Budget for the MegaMan Development Project

Project name	Labor budget	Availability premium	Total budget
MegaMan Development Project	$1,067,600	$33,300	$4,178,916

10. **The earned value schedule.**

Earned value is a very simple, yet very misunderstood, technique of traditional project management. It's the sort of thing that software users and business school students don't like to deal with because they are kind of intimidated by it. Needlessly.

A few years ago, I outlined the basic approach of Total Project Control (TPC) to the senior management of an international construction company based in Boston. After listening to a description of such techniques as the TPC Business Case, the DIPP (Devaux's Index of Project Performance), and the VBS (value breakdown structure), one of the executives said: "Well, this is basically just earned value, right?" That told me (and several others in the room) that not only had he not understood TPC, he also did not have a clue about earned value.

Earned value, in a nutshell, is this. Every activity in the project is "weighted" by some attribute that is common to all the activities. Then, as each activity is completed our project is said to have "earned" the value of that predetermined "weight." The attribute used for weighting activities may be anything (man-hours, miles of highway, schedule risk), but the most common "weight" is budgeted dollars. In other words, each activity is assigned the "earned value" equal to its budget for resources. As each activity is completed, the earned value of the project will increase by the amount of dollars originally budgeted for that activity, irrespective of how much it actually costs to perform the activity. In other words, the schedule along which the earned value should mount (Figure 10.6) is *identical* to the cumulative planned cost accrual schedule shown in the histogram in Figure 10.5.

The standard name for the cumulative earned value planned schedule has been adopted from U.S. Department of Defense (DoD) procedures, and is called the Budgeted Cost for Work Scheduled, or BCWS. Several years ago, the Project Management Institute (PMI) decided to simplify by replacing the four-letter acronym with a two-word term and two-letter acronym: planned value and PV. While this change may have made earned value techniques more accessible (its usage has definitely expanded, largely as a result of PMI's efforts), it has also caused confusion. As this book has attempted to make clear, there is a great deal of difference between value and cost, and earned value is about cost, not value.

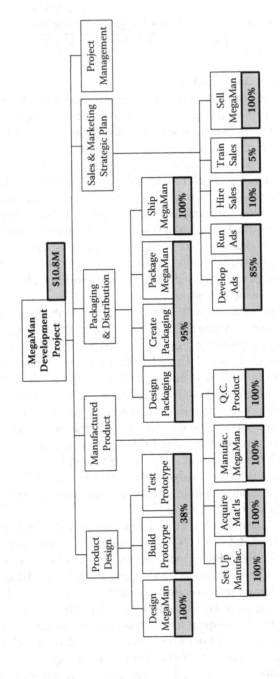

Figure 10.4 VBS for the MegaMan Development Project.

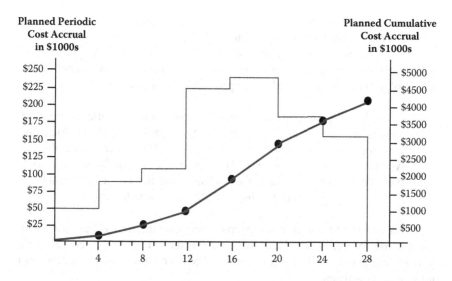

Figure 10.5 Periodic and cumulative planned cost accrual histogram.

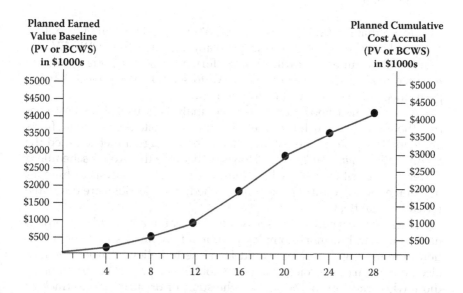

Figure 10.6 Planned value (PV) or budgeted cost for work scheduled (BCWS) (identical with the cumulative planned cost accrual curve shown in Figure 10.5).

The term for the earned value baseline should be planned cost, or PC; that would remove much of the confusion.

When work begins on the project, two types of data get plotted against this one schedule (whether called planned cumulative cost accruals or BCWS or PV or earned value plan). They include:

1. Earned value, equal to the original budgets for all activities completed. This is also called (again from U.S. DoD procedures) the Budgeted Cost for Work Performed, or BCWP (EV in PMI's new terminology).
2. Accrued cost, or the amount of money that was *actually* spent to complete those activities (aka the Actual Cost for Work Performed, or ACWP or AC for Actual Cost in PMI's new terminology).

Those three four-letter acronyms have probably scared people away from using the very useful and quite simple technique of earned value. So here they are listed with their two-letter PMI acronyms and their common language meanings:

BCWS (PV) = Schedule of planned cumulative cost accruals (as in Figure 10.6).
BCWP (EV) = Earned value achieved, or the budgets of the activities accomplished as of a given data date.
ACWP (AC) = Actual costs on the project as accrued to the current data date.

Notice that the first of these, the BCWS-PV (as shown in Figure 10.6), is the only one assembled during the planning stage; the BCWP-EV and ACWP-AC accumulate during the performance of the project and are reported against the curve of the BCWS to see how the project is being performed compared to what was planned.

As mentioned above, earned value analysis is useful for cost control, though not so much for schedule control despite the fact that efforts are made to extend its utility into that arena through a metric called the *Schedule Performance Index (SPI)*. However, because the project schedule is driven by the critical path (which most earned value analysis techniques don't take into account), earned value schedule projections are often distorted by total float.

In general, earned value is less complex than many would-be project managers fear, but more complex than its often simplistic implementation in many organizations would lead one to believe. Earned value is also not primarily a project manager's tool; it is a project control tool for those who are funding the project (the sponsor or customer) to track its performance. As the vice president of the engineering division of a major aerospace/defense contractor used to say: "If a project manager needs earned value reports in order to know the project is in trouble, he's in deep trouble anyway."

For those who wish to further explore the topic of earned value management—its benefits, its shortcomings, its detailed functioning, and its advanced metrics—it is covered in depth in two full chapters of my book *Managing Projects as Investments: Earned Value to Business Value* (CRC Press, 2014). I feel that the whole topic fits much better in that book, among the methods and processes that organizations and senior managers might want to implement for more profitable project management and control, than tucked into the back of this book for project management practitioners. If the project is planned and managed correctly, the project control processes will take care of themselves, provided that those who supervise the project understand project management.

Reporting progress

As the project is performed, efforts should be made to stick to or surpass the plan, and the most important "guiding" aspect of the plan is the schedule. Activities should start, as well as finish, as early as possible, but no later than scheduled. If finishing a critical path activity on schedule means using additional resources and, thus, exceeding an activity's budget, it often makes sense to do it rather than to finish late. Again, it usually takes an awful lot of resources to balance out the cost of a week's delay. However, if it means that the budget for the entire project (including any cost reserve) is to be exceeded, then approval for the extra spending must come from the sponsor/customer who is investing in the project.

Project tracking, like project work, must be performed on a delegated basis. A project manager simply does not have time to run around checking all of the work taking place on a major project. Therefore the activity managers must check the work (or, of course, on a very large activity, get reports from *their* subordinates) and report how it's coming against the scope document requirements, the schedule, and the budget.

A reporting schedule of once a month is often used when dealing with customer projects, such as Department of Defense programs for the U.S. government. However, a monthly reporting schedule is totally inadequate for trying to manage a project. The cost of a month's delay is almost invariably six figures, and often seven. If the project is being delayed, we need to identify the problem and resolve it long before it has had such a huge impact. On nuclear plant refueling projects, the cost of each day that the plant is shut down can be worth up to $2 million. Therefore refueling projects are sometimes scheduled with activity estimates measured in quarter-hours, and with schedule updates following each eight-hour shift.

If nuclear plant refueling can be managed in such an online, real-time manner, then the same *could* be done for any project.

However, this is not necessary for most projects. For most product development projects, for example, my recommendation would be that progress be reported by the project manager to senior management no less often than every two weeks, and that the activity managers update the project manager at least once a week.

That would be the regular schedule. Obviously, there also should be a mechanism by which ad hoc updates are both required and submitted. The project manager may be aware that a portion of the project is especially risky, and demand daily updates. If an activity manager suddenly realizes that his activity is going to take three weeks longer than planned, that information must be passed on to the project manager immediately, and the project plan updated to reflect the bad news.

The planned DIPP baseline

Project management, at any level, must depend on management-by-exception. There is simply too much going on, on one project or across the portfolio of projects, to waste time checking on work that is doing just fine. In addition, the time-sensitive nature of a project demands that a problem must not only be identified as early as possible, but also resolved as quickly as possible. Wherever possible, a problem should be dealt with by the person immediately responsible. However, if that person is unable to resolve the issue without either:

- changing the work scope,
- slipping the schedule, or
- exceeding the activity budget by more than a stipulated amount,

then the problem may have to be escalated to the next highest level. This is true whether we are talking about activity to project level, or project to program or portfolio level.

A common management tool is to incorporate procedures that fix a certain *threshold level* on the schedule or cost of the work. It is the job of the project review process, at this last gate, to set the metrics and threshold levels for escalation. In less sophisticated organizations, the threshold level is usually a raw number. The schedule can slip four weeks, or the cost exceed the budget by $200,000, before the project manager is required to submit to a senior management or customer review. If earned value projections are being used, the threshold levels are usually set on those values. Escalation is required if the work performed (BCWP-EV) falls 10 percent behind the work scheduled (BCWS-EV), or if the actual costs for work done thus far are exceeding the budgets by more than 10 percent. (The earned value metrics have the big advantage of identifying potentially disastrous trends long before the catastrophic point has been reached.)

The trouble is that none of these thresholds is really the most important data item. Schedule and cost are only indirectly important, and by inference. What is *really* important is the three-way interaction of scope, schedule, and cost—the expected monetary value of the project, as measured by the DIPP (Devaux's Index of Project Performance). The project's DIPP is what the project manager and his team should be working, at all times, to maximize. It's when the DIPP descends below a certain level, indicating a specific reduction in expected profit, that senior management should get nervous and involved.

As we discussed in Chapter 1, the simple or Tracking DIPP formula is:

DIPP = ($EMV ± $acceleration/delay) ÷ $Cost ETC

Just as a baseline can be set for cost accruals or earned value that we can then track actual performance against, the same thing can be done for the DIPP. At the start of the project, we assume the expected monetary value will remain pretty constant throughout the project, with no expected changes in the market. The only items that we anticipate will impact the project's expected value if finished exactly on the target date are risk/opportunity factors. If a specific risk factor that would negatively impact our EMV by $500,000 has a 20-percent probability of occurring or being resolved and retired at the end of Week 10, then our baseline DIPP should be reduced by 20 percent × $500,000 or $100,000 through the end of Week 10. However, then after Week 10, the baseline DIPP should anticipate the retirement of that risk and jump by $100,000. Otherwise we expect the EMV to remain stable.

We also expect the schedule to remain stable; we anticipate the project to finish on the scheduled target date. This means that we anticipate neither acceleration premium nor delay cost, and thus it is assumed that the numerator of the DIPP formula will remain stable in the baseline.

This is not the case with the denominator of the DIPP. We assume that, as work is preformed, the denominator will not only decrease, but decrease at a forecast rate, as the exact complement of our cumulative cost accrual curve.

For every dollar we expect to spend, we also expect that our Cost ETC (estimate-to-complete) will be reduced by one dollar. Thus when we plot our planned cumulative cost accrual function, we also are plotting (if we only bother to extract the information) our planned Cost ETC function, as shown in Figure 10.7.

If we expect the two items in the numerator of the DIPP to remain unchanged, but expect the denominator to decrease at a predicted rate,

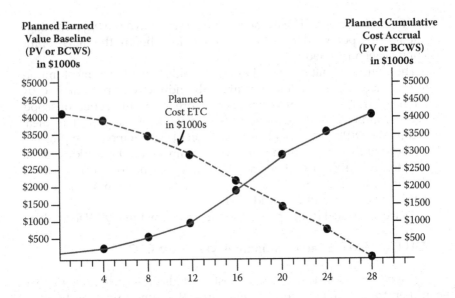

Figure 10.7 Planned cost ETC function plotted as the complement of the cumulative cost accrual function.

then we expect the DIPP to *increase* at a steady and predicted rate. This can allow us to plot the Planned DIPP baseline function as shown in Figure 10.8.

As the last item of work on the project is completed, and the last dollar is spent, the estimate-to-complete becomes 0, and the DIPP rises to infinity. But, long before that, it will have risen far about its original level (or so we most fervently pray). Understand, this is *not* a distortion, or something for which we have to make allowances. It's *reality*. When we have completed half the work of the project, we must no longer factor into our decisions the money we have already spent. Those are *sunk costs*, gone no matter what decision we make. All that we can affect is the future. All that we can save, for example, on the project by canceling it is the money we have yet to spend. So what matters is the ETC; what we have to spend to finish the project and get its EMV. As a project gets closer and closer to completion, it becomes a better and better investment for the remaining funds needed to finish it.

What is needed is a forecast of the periodic DIPP, not dissimilar to the histogram of the accrued costs. For the MegaMan project, the DIPP curve would start at the EMV of $10,800,000 divided by the budget of $4,070,083 = 2.65. It should then rise to the level at Week 28, when the last few hundred thousand dollars or so on the *Package MegaMan, Sell MegaMan,* and *Ship MegaMan* activities are scheduled to accrue. (Note that at the end of Week 14, the Planned DIPP is 3.95.)

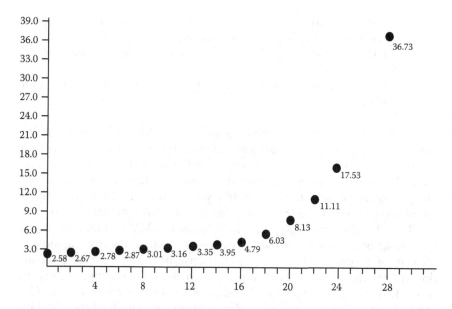

Figure 10.8 Histogram of the Planned DIPP baseline function.

Variances in the actual DIPP

What actually happens to the DIPP during project execution, of course, could be very different.

1. The schedule could slip, causing delay costs to shrink the EMV.
2. The accrued costs could grow, causing trends that show that the Cost ETC will be higher than planned and, thus, force the project over budget to generate less profit.
3. Work scope could be pruned, to preempt either schedule slippage or cost overruns, and result in a less attractive and valuable product.

All three of the above are in the area of the project work, and are the responsibility of the project team and the project manager.

However, there is a fourth reason why the Actual DIPP could shrink below the planned baseline, and the project manager would likely never know because the reason for the DIPP shrinking would be totally unrelated to the project work:

4. External forces, such as the retail market or technological context, could cause the expected monetary value of the final product to shrink. In fact, such external factors can impact the project value in any number of ways:
   a. The delay costs could increase or decrease.

    b. Product features that had been omitted in the original specs could suddenly become very valuable.

    c. Other features that had been included might become of no importance.

    d. Or, the entire market for our forthcoming product could disappear.

All of the items in no. 4 are changes about which the typical project manager, doing a project manager's job, would normally never become aware. Yet they are factors that are crucial to the success of our project. Who *should* be aware of them? The product manager. Or marketing manager. Or sponsor/customer. Or whatever title you want to give to the individual whose job it should be to track the project's expected monetary value.

The TPC Business Case, with its forecast of EMV and its estimate of the impact of different variables on the project plan, is the driving force at the start of the project. The TPC Business Case, however, must not be abandoned once the projects starts; it too is a working document intended to be tracked and updated and modified when changes occur in its data. The product manager/customer should be responsible for checking the data and ensuring their accuracy as often as the project manager is required to report on the project. The product manager and project manager need to work together as closely as possible, with the goal always being to maximize the DIPP and the DIPP Progress Index (DPI, i.e., Actual DIPP ÷ Planned DIPP). If they are not one and the same person (an innovation that has been tried on occasion, although both functions may be too much for any one human being), then perhaps they could be surgically joined at the hip. Whatever the solution, close interaction between these two roles is mandatory.

The DIPP performance index (DPI)

Just as the Planned DIPP can be plotted, the Actual DIPP can be reported and measured on an ongoing basis, along with data regarding actual schedule and cost performance. On this basis, new estimates for project completion date and cost estimate-to-complete (Cost ETC) are plugged into the DIPP formula on the basis of the reported schedule and cost data to estimate the Actual DIPP as of the current data date:

Actual DIPP = ($EMV ± $acceleration/delay) ÷ $Cost ETC

The project can then be monitored on the basis of how closely it is tracking to the Planned DIPP. This, of course, tells senior management and/or the sponsor/customer how the project is progressing in terms of the most important element: expected project profit (EPP).

At 14 weeks, halfway through the project, our accrued costs (from the BCWS curve) should be approximately $1,168,000. Therefore, if we are on schedule and budget at that point, our Cost ETC would be:

Cost ETC = project budget minus accrued costs = $4,070,083 minus $1,335,907 = $2,734,175

Therefore, at the end of 14 weeks, if everything is on target, the project DIPP should be:

DIPP = $10.8 million divided by $2,734,175 = 3.95

If anything has changed to impact either the numerator or the denominator, the Actual DIPP will be different. For example, if everything else is exactly according to plan except that the schedule has slipped by one week, the resulting delay cost of $400,000 would reduce our EMV to $10,400,000. Now the Actual DIPP should be:

Actual DIPP = ($10.8 million – $400,000) divided by $2,734,175 = 3.80

In this case, the DIPP Progress Index would be:

DPI = Actual DIPP divided by Planned DIPP = 3.80 divided by 3.95 = .96

This means that the remaining investment on this project will bring in only 96 percent of the original planned value.

If senior management wants, it can put a threshold on the DPI so that if at any point, due to delays or overspending or changes in the market, the Actual DIPP ever slips below .95 of the Planned DIPP, senior management will immediately be notified. Again, this allows the project to be tracked in the most important terms, its expected value based on the integration of scope, schedule, cost and risk, rather than on tangential tracking of schedule and/or cost in isolation. Any time, at any stage of the project, that the profitability of the project dips below a predetermined floor, alarm bells will sound, signaling the need for senior management intervention.

Working to maximize the DPI

Of course, what people are being tracked on is what they will try to perform best on. If the project team knows that it is being tracked on the basis of the DPI, it will try to maximize the DPI. It will usually do this by trying to speed up the schedule and gain an acceleration premium through earlier projected delivery. And, remember, pruning scope to accelerate won't work because a scope change will change the value breakdown structure and reduce the project's EMV.

All of this will cause the team to look for opportunities for acceleration on the critical path, such as drag identification and, perhaps, working weekends or longer hours to finish one activity so that its critical path successor can start the next week. Suddenly a multiprojected organization will have all of its teams working to maximize the value of the organization's projects.

And wouldn't that be a terrible thing.

chapter eleven

Conclusion

Despite having been around, with most of its techniques and benefits, for more than half a century, traditional project management is still largely ignored. The only exceptions are in those industries and applications where either it is mandated, such as NASA or U.S. Department of Defense projects, or where its value is overwhelmingly clear, such as nuclear power and refinery outages where the delay cost that can be computed easily is often in excess of $1 million a day. However, the vast majority of the business world still either ignores the techniques or uses them in an informal and nonsystematic way that drains the wholistic methodology of much of its potential benefit. (As one senior manager said once during a client conference: "What we really need here as project managers is a bunch of Alpha male dogs." The facial expression of the only woman present caused some of the other men in the room to add hurriedly: "Or Alpha female dogs." But the message was clear—methodology was considered less important than testosterone.)

So if even traditional project management has struggled to establish itself in the corporate world, why should Total Project Control (TPC) have any better luck? The answer is because TPC is tied to what is *really* important to the organization: value and profit. Even in nonprofit organizations, there are divergent values among the various projects. How should one judge between them? The resources that are assigned to them are paid for in dollars. If we decide to pay more for resources on one project than another, aren't we automatically suggesting that one is "worth" more dollars than the other? No matter what the value system is that generates the two projects?

Everything in life requires prioritization. Getting the most benefit from resource usage is the goal, even if that benefit is measured not in profits, but in reduced infant mortality rates or elderly people made comfortable or numbers of distant stars examined for radio signals. The fewer resources we can use to do good, the more we have left over to do better. By monetizing project benefits, we can make decisions about the targeting of our resources, at the program, the project, and even the activity level.

In every organization I have worked with in the past decade that has tried to implement project management, someone invariably makes the point: "In order for this to work, senior management has to understand it and get behind it." They are absolutely right, but, more often than not, executives regard project management as something to be done by their underlings. After all, they never studied this stuff in their MBA programs (and in the few cases where they did, it was with professors who also didn't have much of an understanding of project management's detailed methods), so the topic is often seen as being beneath the notice of a professional executive with an MBA.

TPC changes all that by providing a way to tie the techniques of project management directly and visibly to what matters to executives; the bottom line that is used to determine how *they* are doing *their* jobs. If the relationship can be clearly drawn between, on the one hand, a programmer getting a $7,500 bonus for working 80 hours a week for two weeks, and an extra $992,500 in profits, *that* will get their attention. If they can leave a computer program running all night, and next morning it has generated a multiproject schedule that offers an extra $2.7 million in expected portfolio profit, *that's* worth buying into. Maybe mandating a standardized project methodology, with project managers being required to produce a work breakdown structure (WBS) and critical path method (CPM) schedule, and functional departments being forced (and given the necessary resources) to maintain up-to-date resource libraries, isn't beneath the notice of senior executives, after all.

The time has come to change the way that corporations are organized and run. If 90 percent of a company's revenues are generated by projects, then that company's business *is* project management. The entire management structure and personnel should be steeped in the project management methodology. Whatever steps are needed to manage projects should be taken: procedures, software, and organizational hierarchies. Steps also should be taken to eliminate whatever interferes with the ability to manage projects.

Business schools, too, are simply going to have to alter their curriculum. Much of what they teach is far less relevant than would be the rudiments of the critical path method. Yet students at the best business schools in the United States continue to obtain master's degrees without being required to take so much as a single course in project management. The fact that business school professors (outside of specific degree programs in project management) often don't understand the critical path method, with all its functionality and benefits, is a problem. Business schools and their faculty would much rather teach the snazzy stuff. You know, the sorts of things that future executives of Fortune 500 companies, the alumni who are likely to make big donations, like to play around with. However, when the relationship between scheduling a project and making a profit is demonstrated to be inextricably close, it's amazing how glamorous project management suddenly becomes.

Within recent years, many universities have started offering graduate degrees in project management. Others offer a project management specialization as part of their MBA program. This trend will continue with greater velocity. It will have to; the alternative is a higher bankruptcy rate. Those corporations that are first to make significant improvement in their management of projects are simply going to put the competition out of business.

A final word about project management software. Back in the 1990s, I had a conversation with the president of the company that makes one of the better and more sophisticated project management software packages. He observed that in the decade or so that he had been in business, project management sophistication had actually decreased. At the time I felt he was right, having noticed exactly the same trend myself. There were more people doing project management than a decade earlier, but it was hard to say that the median skill level had improved.

However, thanks in no small part to the diligent efforts of the Project Management Institute (PMI) and the International Project Management Association (IPMA), the depth of understanding of two fundamental techniques has improved far beyond what it was when the first edition of *Total Project Control* was published in 1999: work breakdown structures and earned value. I have hopes that the same thing will happen soon in two other areas: critical path analysis and managing projects for the specific benefits/value they are intended to generate.

Things are getting better, and it is true that a march of a thousand miles starts with a single step. So, project management practitioner, what will be your first step in implementing some of these methods where you work?

Glossary

ABCP *See* **As Built Critical Path**.

AC *See* **Actual Cost**.

Acceleration Premium *See also* **Delay Cost**. Increase in the value of the final product due to earlier completion and/or delivery date.

Actual Cost (AC) The realized cost incurred for the work performed on an activity during a specific time period. Referred to by the U.S. Department of Defense as Actual Cost for Work Performed (ACWP).

Ancestor All activities that precede another activity on a logical path, including immediate predecessors, their predecessors, their predecessors' predecessors, etc.

As Built Critical Path (ABCP) The actual critical path on the final schedule representing the completed work as performed. This is the critical path consisting of work, delays, and constraints that ultimately determines the final length of the project.

Availability Premium A cost to a project, program, or organization to guarantee that a specific resource will be available when needed. This can take the form of overstaffing or paying a fee to reserve the specified resource.

Backward Pass Algorithm *See* **CPM Algorithm**.

BCWP *See* **Earned Value**.

Beta Distribution A Gaussian distribution of probability used in three-point estimating and Monte Carlo simulations to predict probabilities of project and activity duration, effort, and cost.

Black Holes Departments or other functional areas that are so under-staffed and overworked that not even information, such as estimates, can escape.

Bottleneck A resource over-allocation during a period of time that makes the resource unavailable to a project and may delay project completion.

Budget Reserve Also called *cost reserve*, this is additional funding available to a project to mitigate risk. Budget reserve and schedule reserve are two different types of management reserve and are the property of the project manager and/or the customer/sponsor to be allotted as they deem appropriate.

Budgeted Cost for Work Performed (BCWP) The term used by the U.S. Department of Defense for earned value. See **Earned Value (EV)**.

Budgeted Cost for Work Scheduled (BCWS) The term used by the U.S. Department of Defense for **Planned Value (PV)**.

Burn Rate The rate at which a project or program is spending money per calendar unit or reporting period.

Burst Point The point in a schedule where one activity precedes two or more successors.

Business Value The value to an organization, program, or project that any work effort is undertaken to generate.

CLUB *See* **Cost of Leveling with Unresolved Bottlenecks**.

Complex Dependencies Schedule relationships other than simple finish-to-start (FS). These can include start-to-start (SS), finish-to-finish (FF), and start-to-finish (SF) relationships, as well as lags.

Cost Estimate-at-Completion (Cost EAC) An estimate of what the project will have cost when it is completed. At the start of a project, its Cost EAC is its budget. As the project progresses, the Cost EAC is often computed by dividing the budget by the cost performance index (CPI).

Cost Estimate-to-Complete (Cost ETC) An estimate of what it will cost to complete a project from any given point. This is often computed by subtracting an ongoing project's actual cost from its cost estimate-at-completion (Cost EAC).

Cost of Leveling with Unresolved Bottlenecks (CLUB) The cost to a project or program in reduced expected project profit (EPP) due to the delay caused by a specific resource bottleneck.

Cost Performance Index (CPI) This is an earned value metric used to perform cost trend analysis. It is a measure of the cost efficiency of budgeted resources expressed as the ratio of earned value to actual cost. The earned value formula for computing it is

CPI = EV ÷ AC (or in U.S. Department of Defense terms, CPI = BCWP ÷ ACWP).

In simplest terms, future over- or underspending is projected to follow the trend of what has happened so far, so that the total budget is divided by the CPI to estimate the Cost Estimate-at-Completion (Cost EAC).

Cost Plus Contract A category of contract that involves payment to the contractor for all legitimate actual costs incurred for the completed work. Sometimes a fixed fee or an incentive based on the contractor meeting specific goals is added to the cost.

Cost Reserve *See* **Budget Reserve.**

Cost/Schedule Integration A system of techniques or software whereby the impact of cost modifications can be seen on the schedule and the impact of any schedule modifications can be seen on the cost.

Cost Variance (CV) The difference between what was budgeted for the work performed to any given point and what it actually cost. The earned value formula is: $CV = EV - AC$ (or in U.S. Department of Defense terms, $CV = BCWP - ACWP$).

CPI *See* **Cost Performance Index.**

CPM Algorithm An algorithm in project management software packages that uses the sequence and durations of activities to compute the possible dates for each activity and to identify the longest (i.e., critical) path. The CPM algorithm actually consists of two separate algorithms: the forward pass algorithm that computes the earliest possible start and finish for each activity, and the backward pass algorithm that computes the latest possible start and finish for each activity without delaying the end of the project.

CPM Finish Date The earliest date that the project can finish based on the calculations of the CPM algorithm.

CPM Schedule The schedule for all activities generated by the CPM algorithm.

CPM Scheduling Generating a schedule based on the logical sequence and work of each activity.

Critical Path The sequence of activities that represents the longest path through a project, and which, therefore, determines the shortest possible duration. A program also often has a critical path, but usually comprised of projects and based on value generation rather than physical logic.

Critical Path Drag The amount of time by which each item on the critical path (activity, constraint, or bottleneck) is delaying the end of the project or, alternatively, the amount of time by which the project schedule could be compressed by reducing the duration of any critical path item to zero.

CV *See* **Cost Variance.**

Deadline Originally, a *real* deadline—the line drawn approximately 20 feet inside the fence of a U.S. Civil War prison camp. Any prisoner venturing beyond that line would be shot by the guards. Now used metaphorically to describe any completion date constraint on a project.

Delay Cost *See also* **Acceleration Premium.** Decrease in the value of the final product due to later completion and/or delivery date.

Dependency *See* **Predecessor.**

Devaux's Index of Project Performance (DIPP) Originally an index for making decisions about when to abort projects, first published in the author's article, "When the DIPP Dips," in the September-October 1992 issue of *Project Management Journal*. In the first edition of this book, the DIPP was simplified to use as a tracking metric known as the Simple DIPP or the Tracking DIPP. That formula plans a baseline against which to track project progress in investment terms:

Simple DIPP = ($EMV ± $acceleration premium/delay cost) ÷ $Cost ETC.

DIPP Progress Index (DPI) A project investment metric that tracks project progress against planned value: DPI = Actual DIPP ÷ Planned DIPP.

Doubled Resource Estimated Duration (DRED) The DRED is a secondary duration estimate based on how long an activity would take if its resources were doubled. Used along with drag cost, it can be a useful tool for reducing an activity's true cost. *See* **Resource Elasticity.**

DPI *See* **DIPP Progress Index.**

Drag Cost The cost in reduced expected project profit due to the time that a critical path item is adding to the project duration.

DRED *See* **Doubled Resource Estimated Duration.**

Early Dates The earliest dates that activities can start and finish based on CPM calculations.

Early Finish (EF) The earliest that an activity can finish based on CPM calculations.

Early Start (ES) The earliest that an activity can start based on CPM calculations.

Earned Value (EV) The measure of work performed expressed in terms of the budget authorized for that work.

Earned Value Baseline *See* **Planned Value (PV).**

Effort The number of labor units required to complete a scheduled activity or work breakdown structure component, often expressed in hours, days, or weeks.

EMV *See* **Expected Monetary Value.**

Enabler Project A project part of whose value comes from enabling another project (often but not necessarily in the same program) to produce greater value.

EVM *See* **Earned Value Management.**

Exempt Employee An employee that does not have to be paid for working overtime.

Expected Monetary Value (EMV) This is the whole purpose for which a sponsor/customer funds a project. It is the monetized business value that a project is expected to generate if it includes specific scope and finishes on a specific date. Factors beyond the responsibility of the project team can cause a project never to achieve its expected monetary value. However, the EMV should be a key metric in all project-based decisions. (Note that EMV is *not* EVM, which stands for earned value management and is a technique for analyzing project cost, *not* project value.)

Expected Project Profit (EPP) The expected monetary value of a project minus its cost.

Externality In economics, an externality is a cost or benefit that is not included in overall measurements and that may affect a party who had no say in its inclusion. For purposes of measurement, the impact of an externality is usually considered to be zero. In project management, the value/cost of time is often left as an externality.

Fifth Edition of the *PMBOK® Guide The Guide to the Project Management Body of Knowledge* is published by the Project Management Institute (PMI) and is considered the authoritative source of current practice in the project management discipline. The fifth edition was published in 2013.

Fixed Price Contract An agreement that sets the fee that will be paid for a defined scope of work regardless of the cost or effort to deliver it. A fixed price contract also can have an incentive fee that allows the contractor to earn an additional amount by meeting or surpassing certain criteria.

Float (slack) The amount of time that an activity's schedule can slip. *See* **Total Float** and **Free Float.**

Follow-on Value Value of a project that may come from other work or projects that it has enabled, such as revenues from a customization project for a system that was developed and sold to a client.

Free Float (Free Slack) The amount of time that an activity can slip without delaying the schedule of another activity.

Functional Department A department in an organization that is charged with certain functions and whose personnel have certain skills, e.g., mechanical engineering or documentation writing.

Functional Manager The head of a functional department who is usually charged with all personnel matters related to that department including assigning individuals to specific projects.

Gantt Chart Perhaps the most ubiquitous project management format, the Gantt chart was developed by Henry Lawrence Gantt at the Philadelphia Naval shipyard in the early 1900s. The date ribbon allows the user easily to see all work that is occurring simultaneously and resources that may be over-allocated (bottlenecks).

Hard Dependency This is a sequencing relationship between two items of work that is forced by the laws of physics, e.g., the wall cannot be papered until it's plastered and the software code cannot be debugged until it's written. It is very difficult for most project managers to find workarounds to the laws of physics.

Investment The outlay of money for the generation of greater value, usually, but not always, for revenue or profit.

Kahneman, Daniel 2002 Nobel laureate in economics. His book, *Thinking, Fast and Slow* (Farrar, Straus, and Giroux, 2013), has significant implications for managing projects and explanations for why such management is often performed poorly.

Lag In CPM scheduling, a planned delay between any two events.

Late Finish (LF) In CPM scheduling, the latest that any activity can finish without delaying the end of the project.

Late Start (LS) In CPM scheduling, the latest that any activity can start without delaying the end of the project.

Management Reserve A discretionary fund of extra time (Schedule Reserve) and/or money (Budget Reserve) set aside to mitigate risk and that the project manager and sponsor can use.

Mandatory Activities Work activities that must be performed in order for the project to be completed. Mandatory activities have value in the VBS equal to that of the entire project as the project cannot be completed without them. *See also* **Optional Activities.**

Marching Army Costs Costs associated with supporting the project for as long as it takes to complete it. These can include project support costs, overhead costs, and opportunity costs. Shortening the duration of the project often has the positive byproduct of reducing marching army costs.

Merge Points A point in the schedule where many predecessors merge into one successor. *See also* **Burst Points.**

Monte Carlo Simulations A technique used to improve cost and schedule estimates by running probabilistic inputs with specific distribution functions thousands of times in order to see the likelihood of any one result.

Net Present Value (NPV) This is an estimate of how much value an investment will generate, taking into account risk and time factors. In this book, the term *expected monetary value* (EMV) is used as an alternative in order to clearly express the fact that the sponsor/customer of a project expects to achieve a certain value.

Net Value-Added The value of an activity or work package in a project taking into account the value it's expected to add minus its true cost.

Opportunity A project risk with a positive potential outcome.

Optional Activity An activity in a project that is not mandatory and that could be left out without completely destroying the value of the project investment. An optional activity has value equal to the value of the project with that activity included minus the value of the project if it were omitted.

Order-to-Market In many industries, whether a product is first, second, third, or later to market is a major determinant of eventual market sales. This can have a major impact on a project's value/cost of time as an acceleration premium or delay cost.

Pacing the Project Performing noncritical activities so that all paths finish at the same time as the critical path.

Pareto Chart A type of chart developed by Vilfredo Pareto and often used in quality analysis to show in descending order the cost of different factors in an organization. This can be a very useful technique for prioritizing the cost of resource bottlenecks.

Parkinson's Law The principle that "work expands so as to fill the time available for its completion."

Percent Complete A way of estimating progress on activities that suffers from being too subjective. To correct this, earned value tracking often relies on activity-driven milestones, which do not require percent complete estimates.

Perfectly Resource Elastic The quality of a small percentage of activities for which doubling the assigned resources will cut the duration to precisely 50 percent of what it was.

PERT This is an acronym that stands for program evaluation and review technique. It was developed by consultants at the consulting firm Booz Allen Hamilton in 1958 for use on the U.S. Navy's Polaris missile program. It uses a weighted formula for estimating activity effort and durations using three estimates for each activity: optimistic, most likely, and pessimistic.

PERT Chart This terminology is used today simply to mean what is more properly called a project network logic diagram. When someone asks to see a PERT chart, they are not usually expecting use of the PERT formula.

Planned Cost A cumulative function within a project based on the budget accruals for the activities. In earned value tracking, this function is what is termed the planned value (PV) function, or the earned value baseline. In U.S. Department of Defense terminology, this would be the budgeted cost for work scheduled (BCWS).

Planned Cost Estimate-to-Complete (ETC) What it is expected to cost to complete the project from any date forward during the project. At the start of the project, the Planned Cost ETC is identical to the budget. However, as the scheduled work is performed and paid for, the Planned Cost ETC should decline by the amount that was budgeted for the work already performed (i.e., as the complement of the planned value function).

Planned Value (PV) The baseline for earned value tracking, it is important to remember that it is a cost function, not a value function. It is the cumulative budgets for all activities as scheduled. In Department of Defense terminology, it is called the budgeted cost for work scheduled (BCWS).

PMI *See* **Project Management Institute**.

Predecessors Also called logical dependencies, these are the activities that come immediately before other activities. They are the most proximate of an activity's ancestors.

Product Scope The features and functions that characterize the product, service, or result of a project.

Program A group of related projects, subprograms, and program activities managed in a coordinated way to obtain benefits not available from managing them individually.

Project In the *PMBOK Guide* terminology, a temporary endeavor to create a unique product, service, or result. This definition unfortunately does not include the vital information that every project is an investment. Perhaps a better definition would be: "An investment in work to create a unique product, service, or result."

Project Business Case This should be part of the documentation during the initiation phase of a project. It should lay out the reason for the project, what its value is expected to be, and how that value may be impacted by factors such as project duration.

Project Control The processes by which a sponsor/customer or other stakeholder can ensure that a project is being performed in a way that will achieve its important goals. Since a project is an investment, the most important of these goals that should be controlled is value above-cost, our project profit.

Project Management Institute (PMI) An organization of professionals in the project management discipline. It is based in the United States, but has chapters throughout the world. It is responsible for several certification programs in the discipline and for publishing periodic updates to the *Guide to the Project Management Body of Knowledge*.

Project Management Journal The refereed magazine of the Project Management Institute (PMI). It was in this journal that the author's article, "When the DIPP Dips," introduced the concept of the DIPP in September–October 1992.

Project Profit The difference between a project's cost and its value. Because every project is an investment, projects are only funded if they are expected to produce a profit in this sense.

Project Scope The work performed to deliver a product, service, or result with the specified features and functions. Sometimes referred to as "work scope."

Project Sponsor The individual who provides the resources to do a project. The sponsor also can be considered the project investor and is usually the person who expects to benefit from the value of the project.

Quality The degree to which a set of inherent characteristics fulfills requirements.

RAD *See* **Resource Availability Drag.**

Remaining Duration The amount of time necessary to finish an activity that has already started.

Resource Availability Drag (RAD) The delay in the project schedule due to a resource bottleneck or other unavailability of a specific resource when it is needed.

Resource Availability Drag Cost The cost in reduced expected project profit due to a schedule delay caused by the lack of availability of a specific resource. *See also* **Cost of Leveling with Unresolved Bottlenecks (CLUB).**

Resource Elasticity The tendency of an activity's duration to expand or contract in response to increases and decreases in resource levels. *See also* **Doubled Resource Estimated Duration (DRED).**

Resource Leveling The process, supported by many project management software packages, to even out resource usage across a project schedule. The two key resource leveling processes are **Time-Limited Resource Leveling** and **Resource-Limited Resource Leveling.**

Resource Library The database of all the resources available to projects in an organization and their usage and availability shown on a calendar.

Resource Over-Allocations *See* **Bottlenecks.**

Resource-Limited Resource Leveling The process to even out resource usage and limit it to the availability of resources as reflected in the resource library. This process often requires delaying the scheduled project completion to live within the limits of resource availability.

Resource-Limited Resource Schedule The schedule produced by resource-limited resource leveling.

Risk Retirement The process of recognizing that an identified risk factor is no longer relevant. A project plan should include dates by which, if a risk factor has not become manifest, the risk and any mitigating tactics associated with it will be retired and no longer impact the project's expected monetary value.

S-curves The three tracking functions of earned value (PV, EV, and AC) as plotted on a schedule.

Schedule Performance Index (SPI) An earned value metric used to perform schedule trend analysis: SPI = EV ÷ PV. The schedule performance index suffers from the fact that, as usually implemented, it does not recognize the critical path or its crucial importance to the project schedule.

Schedule Reserve Management reserve in the form of extra time in the schedule.

Schedule Variance (SV) Schedule variance is simply the difference between what was budgeted for the work that has actually been completed (EV) and what was budgeted for the work that was scheduled to have been completed so far (PV). The formula is: SV = EV − PV. Like the schedule performance index, it suffers from the fact that the way it is usually implemented does not recognize the critical path or its crucial importance to the project schedule.

Simple DIPP A simplified version of the DIPP as originally published in *Project Management Journal* that can be created as a baseline and tracked during project execution. The formula for the Simple DIPP is: ($EMV ± $acceleration premium/delay cost) ÷ $Cost ETC. Also known as the Tracking DIPP.

Slack *See* **Float**.

Soft Dependency Also known as a discretionary dependency, this is a predecessor–successor relationship input to the schedule on the basis of a project manager's decision rather than on the basis of the logic of the physical work. If the soft dependency is on the critical path, then the drag and drag cost should be computed to ensure that such a dependency is not costing more than it's worth.

SPI *See* **Schedule Performance Index**.

Sponsor/Customer The individual, individuals, or organizations that initiate a project and invest in it and that hope to reap the value that it generates.

Successor Activity An activity that comes immediately after another activity. Successor activities are the most proximate descendent activities to any given predecessor.

Thinking, Fast and Slow A book (Farrar, Straus, and Giroux, 2013) by 2002 Nobel laureate Daniel Kahneman that may explain many of the reasons that many project management decisions are made that are unjustified from an economics point of view.

Time-Limited Resource Leveling The process to even out resource usage and limit it to the availability of resources as reflected in the resource library. In time-limited resource leveling, the project completion date is fixed. If the date that is input is the CPM finish date, then no activity can be delayed beyond its total float and the schedule will continue to reflect resource bottlenecks that will have to be resolved.

Too Many Cooks Syndrome A situation on a project where adding resources causes an activity to take longer. The DRED on such activities is higher than the original estimate.

Total Float (Total Slack) The amount of time that an activity's schedule can slip without delaying the end of the project.

Tracking DIPP *See* **Simple DIPP.**

Triangular Distribution One of the standard distribution shape that a user can select for the durations of all the project activities to be scheduled in a Monte Carlo simulation software package. The triangular distribution, which is based on the input of lowest, most likely, and highest estimates, is the most common distribution shape used. The second most common distribution shape used is what is generally called the beta distribution, which reflects the PERT formula.

True Cost (TC) The true cost of doing work in a project is due both to the cost of resources for that activity and the impact of that activity's duration on the project schedule. The formula is: TC = resource cost + drag cost. The net value-added of an activity is its value-added minus its true cost. Because most projects do not compute the drag cost of activities, they often wind up including work that has a negative net value-added.

Utilization Rate This is the percentage of time that a resource is billable to a specific customer or contract.

Value Breakdown Structure (VBS) A concept introduced in the first edition of this book that separates a project's work packages and activities into mandatory and optional categories and estimates the value-added for each optional activity.

Value Drivers The factors that drive the business value from any given project. Expected revenues and savings are the two most easily identified value drivers, but others, such as follow-on business, customer satisfaction, patents, market dominance, and giving engineers experience with valuable new technologies, also can drive significant value on certain projects.

Value-Added The difference between the value of a project with or without a specific work package or activity.

VBS *See* **Value Breakdown Structure.**

Appendix

Exercise A

The following is a list of nine project activities, with their durations in days. (All relationships are finish-to-start with no lag.)

Activity	Duration	Pred.
A	10D	None
B	15D	A
C	13D	A
D	16D	B
E	7D	B, C
F	8D	C
G	4D	D
H	5D	E, F
I	5D	G, H

1. How many days will the project take to complete?
2. Which activities comprise the critical path?
3. What are the total float and critical path drag of each activity?
4. If the expected monetary value of the project is $1 million if completed in 48 days, with a delay cost of $100,000 per each day later and

an acceleration premium of $50,000 per each day earlier, what is the drag cost of each critical path activity with the current schedule?

5. If the project budget is $400,000, what is the starting DIPP with the current schedule?

6. If Activity E takes 20 days instead of 7, how much longer or shorter would the project be?

7. How much more or less would the project's EMV (expected monetary value) be than with its current schedule if Activity E takes more than 20 days?

8. If Activity D takes 6 days instead of 16, how much longer or shorter would the project be?

9. How much more or less would the project's EMV be than with its current schedule if Activity D takes 6 days?

10. If Activity G has a value-added of $450,000 and its resources cost $75,000, how much is its net value-added?

Exercise B

The following is a list of eight project activities, with their durations in days. Each activity's predecessor(s) are listed along with type of relationship and lag amount if any.

Activity	Duration	Pred.	Relationship	Lag
A	8D	None	–	–
B	16D	A	Start-to-start	5
C	12D	A	Finish-to-start	2
D	17D	A	Start-to-start	4
E	12D	B, C	All finish-to-start	0
F	4D	C, D	All finish-to-start	0
G	6D	F	Finish-to-start	0
H	5D	E, G	All finish-to-start	0

11. How many days will the project take to complete?

12. Which activities comprise the critical path?

13. What are the total float and critical path drag of each activity?

14. If the expected monetary value of the project is $1 million if completed in 38 days, with a delay cost of $100,000 per each day later and an acceleration premium of $50,000 per each day earlier, what is the drag cost of each critical path activity with the current schedule?

15. If the project budget is $400,000, what is the starting DIPP with the current schedule?

16. If Activity D takes 22 days instead of 17, how much longer or shorter would the project be?

17. How much more or less would the project's EMV be than with its current schedule if Activity D takes 22 days?

18. If Activity C takes 10 days instead of 12, how much longer or shorter would the project be?

19. How much more or less would the project's EMV be than with its current schedule if Activity C takes 10 days?

20. If Activity E has a value-added of $200,000 and its resources cost $75,000, how much is its net value-added?

Answers to Exercise A

1. How many days will the project take to complete? **50 days**
2. Which activities comprise the critical path? **A-B-D-G-I**
3. What are the total float or critical path drag of each activity?

Total float
C = 9, E = 8, F = 9, H = 8
Drag: A = 10, B = 9, D = 8, G = 4, I = 5

4. If the expected monetary value of the project is $1 million if completed in 48 days, with a delay cost of $100,000 per each day later and an acceleration premium of $50,000 per each day earlier, what is the drag cost of each critical path activity with the current schedule?

Drag cost: A = $600,000, B = $550,000, D = $500,000, G = $300,000, I = $350,000

5. If the project budget is $400,000, what is the starting DIPP with the current schedule?

($1 million – $200,000) ÷ $400,000 = 2.0

6. If Activity E takes 20 days instead of 7, how much longer or shorter would the project be? **5 days longer**

7. How much more or less would the project's EMV be than with its current schedule if Activity E takes 20 days?

$500,000 less

8. If Activity D takes 6 days instead of 16, how much longer or shorter would the project be?

8 days shorter

9. How much more or less would the project's EMV be than with its current schedule if Activity D takes 6 days?

$500,000 more

10. If Activity G has a value-added of $450,000 and its resources cost $75,000, how much is its net value-added?

$75,000

Answers to Exercise B

11. How many days will the project take to complete? **39 days**
12. Which activities comprise the critical path? **A-C-E-H**
13. What are the total float or critical path drag of each activity?

Total float: B = 1, D = 3, F = 2, G = 2
Drag: A = 6, C = 1, E = 2, H = 5 (A-C lag = 1)

14. If the expected monetary value of the project is $1 million if completed in 38 days, with a delay cost of $100,000 per each day later and an acceleration premium of $50,000 per each day earlier, what is the drag cost of each critical path activity with the current schedule?

Drag cost: A = $350,000, C = $100,000, E = $150,000, H = $300,000.

15. If the project budget is $400,000, what is the starting DIPP with the current schedule?

($1 million – $100,000) ÷ $400,000 = 2.5

16. If Activity D takes 22 days instead of 17, how much longer or shorter would the project be?

2 days longer

17. How much more or less would the project's EMV be than with its current schedule if Activity D takes 22 days?

$200,000 less

18. If Activity C takes 10 days instead of 12, how much longer or shorter would the project be?

1 day shorter

19. How much more or less would the project's EMV be than with its current schedule if Activity C takes 10 days?

$100,000 more

20. If Activity E has a value-added of $200,000 and its resources cost $75,000, how much is its net value-added?

Minus $25,000

Index

Printed in the United States
by Baker & Taylor Publisher Services